Sustainable Cloud and Energy Services

Wilson Rivera
Editor

Sustainable Cloud and Energy Services

Principles and Practice

Springer

Editor
Wilson Rivera
School of Engineering
University of Puerto Rico at Mayaguez
Mayaguez, PR, USA

ISBN 978-3-319-62237-8 ISBN 978-3-319-62238-5 (eBook)
DOI 10.1007/978-3-319-62238-5

Library of Congress Control Number: 2017953011

Printed on acid-free paper

This Springer imprint is published by Springer Nature
The registered company is Springer International Publishing AG
The registered company address is: Gewerbestrasse 11, 6330 Cham, Switzerland

Preface

Cloud computing services support sustainability in different ways. For example, the design of more efficient cloud services can contribute in reducing the energy consumption and environmental impact by providing low-cost access to information and services. Similarly, cloud computing services provide citizens, especially those living in vulnerable or underserved communities, a platform to become key actors in a sustainable energy market. In such cloud-based platform, energy services can be dynamically combined to establish service dependencies between energy producers and consumers with the goals of reducing costs, reducing pollution, and bringing electricity to more citizens.

The chapters in this book will address conceptual principles and illustrate the latest achievements and development updates concerning sustainable cloud and energy services. This book will serve as a useful reference for advanced undergraduate students, graduate students, and practitioners interested in the design, implementation, and deployment of sustainable cloud-based energy services. The book might be of interest for professionals in the areas of power engineering, computer science, and environmental science.

The distinguishing features of this book are as follows:

- It provides a state-of-the-art look at sustainable cloud-based energy systems.
- It brings together contributions from distinguished researchers and experts in the area.
- It presents both well-established and novel techniques for designing, developing, and deploying sustainable systems.
- It provides a set of sample projects in this area with realistic constraints.

In Chap. 1, the authors provide a comprehensive survey of cloud computing and Internet of things (IoT) technologies. The authors describe the integration of cloud computing and IoT domains, the challenges of such integration as well as the products currently available, and the applications in agriculture, healthcare, and smart cities. This chapter is a very good introduction to cloud computing terminology and concepts.

In Chap. 2, the authors describe a model for decentralized networks of self-government objects that efficiently share electrical power from a common, limited power supply. In the proposed model, each object shares with other objects its priority and power demand by broadcasting messages over the network, and consequently each object decides its own actions based on a fusion of data collected from the network. The authors in this chapter introduce decentralized power distribution (DPD) concepts, power allocation and redistribution algorithms, and communication protocols needed to fully implement such a network. In addition, they provide meaningful proofs of software correctness and simulation results.

In Chap. 3, the authors describe the energy service platform (ESP) which is defined as a software-based product that serves as a foundation on which outside parties can build complementary products or services. ESP is conceived as a cloud-based service platform for the energy domain, applying the service-oriented architecture (SOA) paradigm in order to provide a loosely integrated suite of services. The ESP can coexist and cooperate with existing management systems or gateways, whose services can be offered to ESP users directly or as part of more complex services.

Awareness of privacy and security issues of cloud-based energy systems is of paramount importance. In Chap. 4, the authors discuss methods for privacy-preserving load profile matching where load profiles from utility providers and load profile forecasts from customers are transformed in a distance-preserving embedding in order to find a matching tariff. In this chapter, the authors address the issue of privacy as an aspect relevant to the sustainability of cloud-based energy platforms.

Energy consumption in buildings is responsible for a significant portion of the total energy use and carbon emissions in large cities. In Chap. 5, the authors discuss utilizing cloud computing to provide energy-related services to enhance energy consumption in smart buildings.

Virtual machine (VM) consolidation is a very important topic in the design of energy-efficient dynamic cloud resource management systems. In Chap. 6, the authors classify and critically review virtual machine consolidation algorithms from different viewpoints. With detailed analysis, the authors highlight important research results, compare different methods of VM consolidation, and indicate areas of future work.

In Chap. 7, the authors address the problem of dynamic resource management in large data centers. They proposed an enhanced instance-based learning (EIBL) approach evaluated using Google Cluster Data.

Cloud providers are moving toward using renewable energy sources to partially power their data centers. In Chap. 8, the authors propose a short-term prediction model that uses the previously observed energy levels to train itself and predict the energy level up to 15 min ahead into the future. As a result, the cloud provider can perform online virtual machine migrations with performance close to the optimal offline, which has the full knowledge of the future level of renewable energy.

In Chap. 9, the authors describe a methodology to determine the optimal size of a river-based micro-hydrokinetic pumped-hydro-storage hybrid system. The authors discuss the effect brought by different demand sectors such as residential, commercial, and industrial load on sizing and operation of the proposed hybrid system.

In Chap. 10, the authors discuss the impact brought by different demand sector profiles on the daily operational cost and optimal scheduling of grid-connected photovoltaic systems with bidirectional power flow which is analyzed for the specific case of Bloemfontein in South Africa.

Mayaguez, PR, USA Wilson Rivera, Ph.D.

Acknowledgments

As editor, I would like to express my gratitude to the authors for their outstanding contributions to this book. I am confident that the work presented by the authors will make a forceful contribution to the field of sustainable cloud and energy services.

This book could not have been assembled without the efforts of Springer's editorial and production staff, specifically Amanda Quinn, Brinda Megasyamalan, and Marie Josephine Chandramohan.

Contents

Chapter 1
Cloud Computing and Internet of Things Integration: Architecture, Applications, Issues, and Challenges

Akash Malik and Hari Om

1.1 Introduction

Cloud computing and IoT are two distinct technologies having wide applications in human life. Their acquisition and use is extremely pervasive, making them future of internet. The IoT is the internetworking of physical devices, embedded with electronics, software, sensors, actuators, and network connectivity, that enable these objects to collect and exchange the data. IoT describes a system where the items in physical world, and sensors within or attached to these items, are connected to the Internet via wireless and wired Internet connections. These sensors can use different types of local area connections such as Radio-Frequency Identification (RFID), Near Field Communication (NFC), Wireless Fidelity (Wi-Fi), Bluetooth, and Zigbee. The sensors can also have wide area connectivity such as Global System for Mobile communication (GSM), General Packet Radio Service (GPRS), 3G, and Long Term Evolution (LTE) [1]. IoT mainly divided into three components that enable seamless ubiquitous computing: (1) hardware—made up of actuators, sensors, and embedded hardware for data exchange; (2) middleware—provides data storage and computing applications based on demand for data analytics; and (3) presentation—an easily understandable visualization and interpretation tools that can be accessed on different platforms and build for various applications [2]. Cloud computing is a type of on-demand Internet-based computing model consisting of autonomous, networked IT (hardware and/or software) resources. It may be considered as the delivery of power of computer as needed, information storage, applications, and other tools of information technology by cloud services platform via the internet with pay-per-use pricing [3]. The service providers offer

A. Malik (✉) • H. Om
Department of Computer Science and Engineering, Indian Institute of Technology (Indian School of Mines), Dhanbad, India
e-mail: akashmalik28@gmail.com; hariom4india@gmail.com

W. Rivera (ed.), *Sustainable Cloud and Energy Services*,
DOI 10.1007/978-3-319-62238-5_1

cloud services through the Internet as a set of easy-to-use, scalable, and economical services to interested clients on subscription basis.

The IoT may be specified by the small things of real world that are extensively distributed and have constrained storage capacity as well as processing power. Due to these constraints, it has issues related to reliability, performance, privacy, and security. Additionally, there exists large amount of heterogeneity in data as well as devices and IOT provides a platform to handle all of them. On the contrary, the cloud computing, a mature technology, provides almost unlimited capabilities in terms of storage as well as processing power, and has helped addressing most of the IoT issues. So, a paradigm consisting of cloud and IoT, two interdependent technologies, together is expected to disorder both current and future Internet [4]. The term given for Cloud and IoT integration is *CloudIoT* or *Cloud of Things* (CoTs). Cloud Computing and IoT integration explores the area of IoT capabilities as well as cloud capabilities. Furthermore, new capabilities like data mining, complex analysis, and real life processing in IoT as well as cloud service area will be explored.

1.1.1 Internet of Things

The terminology *Internet of Things* (IoT) was first introduced in 1999 by British technology pioneer Kevin Ashton to describe a network system where the objects in physical world can be connected to the Internet by sensors. Ashton further devised this term to highlight the capability of connecting the RFID tags to the Internet. In IoT, "things" refer to any object on the face of the Earth irrespective it is a communicating device or a non-communicating object. Today, the IoT has become a trendy term in which the Internet connectivity and computing capability are continued to various categories of objects, devices, sensors, and usual items. Despite universal acceptance of IoT, there is no single globally recognized definition for the term. There are various definitions of IoT [1] and are as follows:

> The term Internet of Things belongs to scenarios in which connectivity of the network and computing capability strikes to objects, sensors, and usual items not personal computers, granting these devices to create, interchange, and utilize data with minimum human involvement.

The Oxford Dictionaries definition:

> Internet of Things (noun): The interconnection via the Internet of computing devices embedded in everyday objects, enabling them to send and retrieve data.

The RFID group defines Internet of Things as:

> The worldwide network of interconnected objects uniquely addressable based on standard communication protocols.

1.1.1.1 IoT Devices, Platform, and Services

We can divide IoT devices into two comprehensive classes: wearable and microcontroller/microprocessor driven embedded IoT devices. Some examples according to categories of hardware devices in IoT are given below.

Wearable Devices Google Glass, Samsung Gear 2 neo, Pebble Watch, Misfit Shine, Android wear, etc.

Embedded System-Based Devices Arundio, Intel Galileo, Raspeberry-Pi, Beglebone, etc.

Accessories Relays, Displays, Switches, Actuators, Sensors, etc.

Smart Devices An object is a smart object that can define its own possible interactions. Any object which has a state as well as has certain information joined with that state which can also determine type of connectivity, time span of connectivity, and connectivity protocol are called smart objects. Smart objects in IoT are Global Positioning System (GPS), Global System for Mobile communications GSM, General Packet Radio Service (GPRS), Radio Frequency Identification (RFID), Near Field Communication (NFC), etc.

Platform and Services Some platform and services of IoT are RIOT, Carriots, Lithouse, Sensinode, IFTTT, Arrayent, Alljoyn, ioBridge, Tizen, SalesForce, Axeda, OpenIoT, etc.

1.1.1.2 Layered Architecture of IoT

On the bases of layered architecture, the IoT is divided into four layers: perception/link layer, network/transport layer, middleware layer, and application/service layer, shown in Fig. 1.1 [5]. These layers are described below.

Perception Layer/Link Layer

The Perception Layer/Link Layer is the bottom layer where sensors, RFID, etc., devices are launched into the IoT network to sense the happening in environment and report to sink nodes via network layer. This layer further divided into two layers: edge technology and access gateway layers, which are discussed below.

(a) *Edge technology layer*: This is a hardware layer comprising embedded systems, sensor networks, RFID tags, and many other different sensors. This layer performs many other respective functions such as gathering information from a system or an environment, processing this information, and assisting communication.
(b) *Access gateway layer*: This layer deals with handling of data generated by IoT devices or things. It is primarily accountable for publishing, and subscribing the services provided by the IoT devices, message routing, and hovelling communication between platforms.

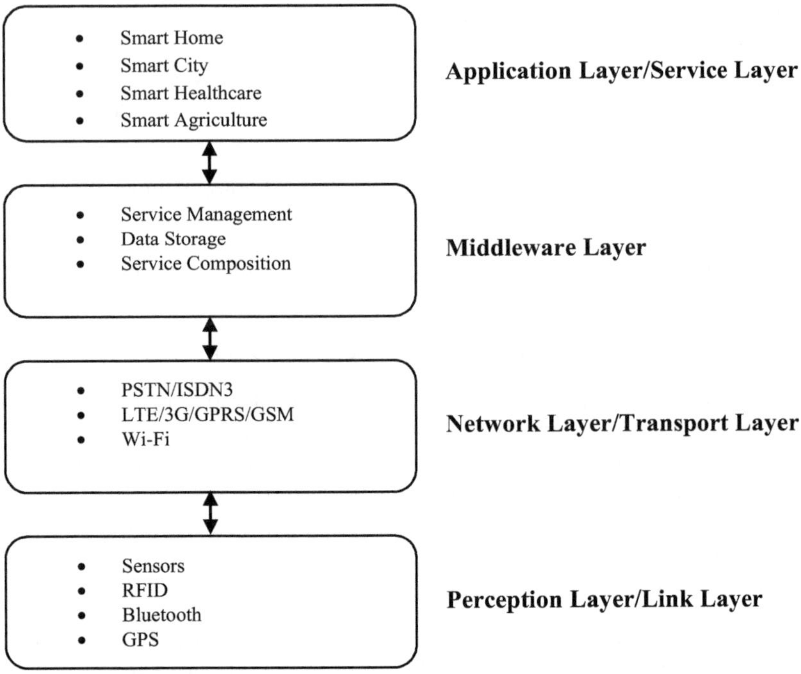

Fig. 1.1 Layered architecture of IoT

Network Layer

This layer may be considered a group of internetworking methods, protocols, and specifications in the Internet protocol suite, which can be utilized to transport datagrams (packets) from a sending host across the network boundaries to a receiving host. The sender and receivers are mentioned by the network addresses (IP addresses) by the Internet Protocol (IP).

Middleware Layer

This layer performs collecting and filtering the data received from the hardware devices, data discovery, and access control to the devices for applications. This layer can be located anywhere like in sensor nodes, sink nodes, high-level application terminal, etc. The main goal of middleware layer is to provide abstraction of functionalities and communication capabilities of the devices in IoT deployment for achieving a ubiquitous integration with other technologies like cloud services [6].

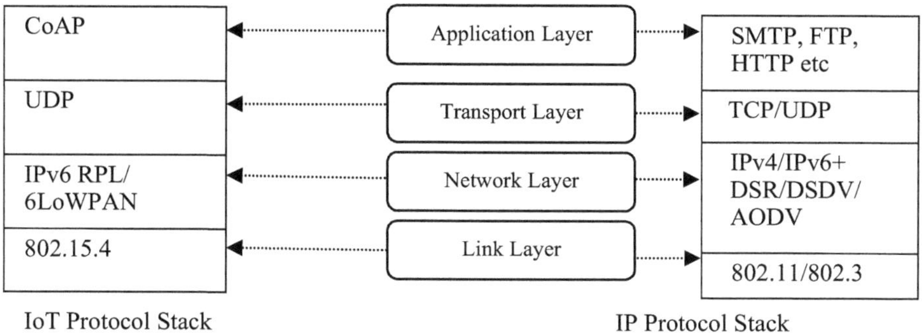

Fig. 1.2 IoT network and traditional IP network

Application Layer

In IoT-based systems, application layer is used for delivering different application services provided over middleware layer to different users and applications. The application services can be used in many industries like healthcare, logistics, retail, etc.

1.1.1.3 Comparison of IoT Network and Traditional IP Network

The IoT devices are resource constrained, i.e., they have limited battery, limited processing power, low transmission range, etc., and also the network links are lossy. The module of traditional IP protocol is heavyweight and hence is not suitable for IoT system architecture. The lightweight protocol corresponding to the traditional IP protocol for resource-constrained devices in IoT has been developed [5, 7]. Figure 1.2 shows the layers in IoT protocol stack vs. IP protocol stack.

(a) *Constrained Application Protocol (CoAP)*: Internet Engineering Task Force (IETF) has developed the Constrained Application Protocol (CoAP), an application protocol for IoT devices, to connect the IoT devices to Internet. It is a replacement of the HTTP protocol for application layer of IoT, considering optimized length of datagram and satisfying the representational state transfer (REST) protocol to support Uniform Resource Identifier (URL). It provides a reliable communication based on user datagram protocol (UDP).
(b) *User Datagram Protocol (UDP)*: The UDP is a lightweight connectionless protocol, suitable for resource-constrained IoT devices, that provides checksum for data integrity and port number to communicate with other nodes. Since the IoT applications are time sensitive, dropping of packets is preferred rather than waiting for delayed packet. The UDP protocol satisfies this requirement and hence has been adopted for transport layer protocol.

(c) *Link Layer Protocol*: The IoT implements IEEE 802.15.4 standard protocol for sensor devices for medium access control (MAC) layer. The frame formats of a traditional network at link layer are not suitable for resource-constrained devices in IoT due to their overhead. The IEEE 802.15.4 defines frame format, header, and communication algorithms for IoT devices that uses channel hopping and time synchronization. Its other features are slotted frame structures, synchronization, channel hopping, network formation, scheduling. Some other MAC layer protocols for IoT applications are IEEE 802.11 AH, wireless HART, Bluetooth low energy, Zigbee smart energy, DASH7, Home plug, G.9959, LTE-A, LoRaWAN, Weightless, etc.
(d) *Network Layer Protocol*: IoT uses IPv6 routing to overcome the address space problem in IoT, and lossy network (RPL) and 6LoWPAN network at network layer for resource-constrained sensor nodes.
(e) *RPL (Routing Protocol for low power and Lossy network)*: The RPL, primarily designed to meet the specific requirement of IP IoT, is a novel standard routing protocol that can perform one-to-one, one-to-many, and many-to-one communication. It supports both unidirectional and bidirectional communication between the root, also called sink node, and constrained nodes for creating a destination oriented directed acyclic graph (DODAG) with the root node. The root node is directly connected to the Internet using IPv6 Boarder Router (6BR) using a three-way handshake as given below.

  - *DODAG information object (DIO) message*: The root node broadcasts the DIO message for the formation of topology.
  - *DODAG advertisement object (DAO) message*: Other nodes select their parents after receiving the DIO messages and reply a DAO message to the parent asking the permission to join it.
  - *DODAG acknowledge message (DAO ACK)*: Parent node permits by sending DAO ACK message based on the rank value calculation for each node using the rank (of possible parent) and other parameters like node energy and node distance. If a new node wants to join a parent, it sends a DODAG Info solicitation (DIS) message to find if any DODAG exists.

(f) *IPv6 over Low Power Wireless Personal Area Network (6LoWPAN)*: The 6LoWPAN integrates Wireless Sensor Networks (WSNs) and IP-based infrastructure such as IEEE 802.15.4 networks and performs context aware header compression as given below:

  - IP Header Compression (IPHC) to compress IPv6 header.
  - Next Header Compression (NHC) to compress the user datagram protocol (UDP) header and IPv6 extension header.

The 6LoWPAN standard redefines fragmentation and reassembly of packets because the payload size of link layer in 6LoWPAN network is very limited and the security protocol for large application data size makes the IEEE 802.15.4 frame size larger than the maximum transmission unit (MTU) size (127 bytes). The 6LoWPAN fragmentation scheme provides reassembly tag and an offset to every fragment and provides additional fragmentation in case the data size exceeds MTU size.

1.1.2 Cloud Computing

Cloud computing, also known as on-demand computing, is an internet-based computing, where a shared pool of resources and information is provided to computers and other devices on demand [3].

The cloud computing has following characteristics: on-demand self-service, broad network access, resource pooling, rapid elasticity, and measured services.

- *On-demand self-service*: A consumer can unilaterally provision computing capabilities such as server time and network storage as per requirement automatically without requiring human interaction.
- *Broad network access*: Various capabilities are available over the network that can be accessed through standard mechanisms to promote use by heterogeneous thin or thick client platforms (e.g., mobile phones, tablets, laptops, and workstations).
- *Resource pooling*: The computing resources such as storage, processing, memory, and network bandwidth are pooled to serve multiple consumers using a multi-tenant model in which the physical and virtual resources of different types are dynamically assigned and reassigned as per demand. There is a sense of location independence in that the customer generally has no control or knowledge over the exact location of the provided resources but may be able to specify location at a higher level of abstraction (e.g., country, state, or datacenter).
- *Rapid elasticity*: It refers to the capabilities that can allocate and release the resources as per demand automatically at any time.
- *Measured service*: Cloud systems automatically control and optimize the resource usage by leveraging a metering capability at some level of abstraction, e.g., storage, processing, bandwidth, and active user accounts. Resource usage can be monitored, controlled, and reported in transparent manner for both the provider and consumer.

1.1.2.1 Services of Cloud Computing

There are three main services provided by cloud computing as given below [3].

(a) *Platform as a Service (PaaS)*: The PaaS is offered as a development environment to an application developer through a toolkit. With Platform as a Service, one (user) deploys self-created or acquired applications onto the cloud infrastructure using the tools and the programming language provided by the provider. The user does not bother about the underlying cloud infrastructure such as network, operating systems, or storage, but he can access and use the deployed applications and the application hosting environment configurations. Google provides this type of service where users create and host the web applications using Google's development.

(b) *Software as a Service (SaaS)*: The SaaS model helps using the applications running on a cloud infrastructure, which can be accessed from different user devices through a thin client interface such as web browser and a program interface. The web-based email and Salesforce, an online sales management, are examples of Software as a Service.
(c) *Infrastructure as a Service (IaaS)*: An IaaS provider provides virtual or physical systems/machines and other resources as per the Internet Engineering Task Force (IETF). In IaaS, one is able to access processing power, storage, and other computing resources. The user does not bother about the underlying cloud infrastructure, but has accessibility over the operating systems, storage, deployed applications, and limited accessibility over certain networking components, such as host firewalls. Amazon provides this type of service where users can rent virtual servers on which they run their own applications.

1.1.2.2 Deployment Models

There are mainly four types of deployment models in cloud environment.

(a) *Private Cloud*: It is the cloud infrastructure provisioned for the use of a single organization with multiple consumers, and hosted either internally or externally, for example, individual business unit in a large corporation. The private cloud may be handled including its ownership and manageability by the organization or a third party, or some of them, and it can be on or off premises.
(b) *Public Cloud*: It is the cloud that provides the services (which may be free) using a network for public use on the premise of the cloud provider, for example, a university.
(c) *Community Cloud*: It is the cloud that is exclusively used by a specific community from organizations having common concerns such as security requirements and policy. It may exist on or off premises and managed by one or more organizations, a third party, or some combination of them.
(d) *Hybrid Cloud*: It is a composition of two or more clouds (private, community, or public), each maintaining its unique entity. However, they together offer collectively the benefits of multiple deployment models.

The remaining chapter is arranged as follows. In Sect. 1.2, the works related to Cloud and IoT integration are discussed. In Sect. 1.3 Cloud and IoT integration architecture components and architecture based on CoAP and 6LowPan are given. In Sect. 1.4 various applications of integration discussed. Section 1.5 points out challenges and open issues. Section 1.6 provides conclusions and future work.

1.2 Related Works

The cloud and IoT have independently seen a rapid evolution, which may be considered complementary to each other and researchers are working on their integration [6, 8–11]. Their integration has storage processing and networking capabilities that mainly come under IoT due to its characteristics. The devices related to IoT contain a lightweight and compatible mechanism for communication with the cloud systems deployed through IoT middleware. Since IoT has devices, technologies, and protocols of different types, its functionalities like scalability, interoperability, reliability, efficiency, etc., are quite challenging to get. Its integration with cloud addresses these types of issues [12, 13] and also provides many other features like ease-of-access, ease-of-use, etc. [12]. Since IoT has limitations in terms of storage, processing, and communication, the cloud computing can help in overcoming these constraints. The cloud can provide an effective solution for IoT service management and implementing applications and services that exploit the things or data produced by them [9]. The cloud can also be benefitted from IoT by increasing its scope to manage real-world things. It will certainly affect the application development in future in which gathering, processing, and transmission of information will cause further issues [10].

The adoption of cloud and IoT integration, which is also termed as CloudIoT paradigm, helps enabling new scenarios for smart services and applications based on the cloud through the things: Sensing as a Service (SaaS) that can provide pervasive access to the user data [14], Sensing and Actuation as a Service (SAaaS) that can provide automatic control logics in a cloud [14], Sensor Event as a Service (SEaaS) that can enable to dispatch messaging services triggered by the sensor events [14], Sensor as a Service (SenaaS) that can provide pervasive management of remote sensors [15], DataBase as a Service (DBaaS) that can provide pervasive database management [15], Data as a Service (DaaS) that can provide pervasive access to any kind of data [15], Ethernet as a Service (EaaS) that can provide pervasive layer-2 connectivity to remote devices [15], Identity and Policy Management as a Service (IPMaaS) than can provide pervasive access to policy and identity management functionalities [15], Video Surveillance as a Service (VSaaS) that can provide pervasive access to recorded video and implementing complex analyses in the cloud [16]. These services have led to many applications such as healthcare, smart home and smart metering, video surveillance, automotive and smart mobility, smart logistics, and environmental monitoring [15]. Deployment of RFID applications in IoT networks is generally complex and costly as they require tedious deployment and management of large and heterogeneous distributed systems [17]. The RFID applications are suitable for large organizations only as small business organizations have limited resources. This issue can be addressed by cloud computing with integration of virtualization technologies and the architecture of web and its services [17]. There are various protocols for integration of cloud and IoT that take care of research oriented and open-source related projects and enterprise products in variety of disciplines. The Nimbits [18], ThingSpeak [19], Paraimpu [20], DeviceCloud

[21], SensorCloud [22], iDigi Device Cloud [23], Stack4Things [24] are some of such protocols. There are some Things development platforms that include Wiring [25], Sun SPOT [26], mbed [27], or Arduino [23].

The architecture of cloud and IoT integration using the combination of Hypertext Transfer Protocol (HTTP) and Message Queuing Telemetry Transport (MQTT) protocol servers has been discussed in [28]. A common integration architecture of cloud and IoT, which is called CloudThings, based on RESTFul web services, and protocols like CoAP, for communication between constrained resources things and constrained networks, and IPv6 over Low power Wireless Personal Area Networks (6LoWPAN) protocol has been discussed in [29]. The extension of cloud, which is closer to things that produce and acts on the data in IoT, is termed as Fog Computing. Integration architecture of Fog, Cloud, and IoT has been discussed in [30]. There are several IoT applications such as e-health, Smart Grid and Power Systems, generating a large amount of data regularly. Therefore, a secure data analytics for cloud and IoT integration (cloud-Integrated Internet of Things) applications is needed. A reference architecture for privacy reservation in cloud-based IoT applications has been discussed in [31] and fully homomorphic encryption systems to achieve data security and privacy over cloud-based analytical phases on three applications: (1) abnormality detection and patient classification in electronic health records, (2) anomaly detection services for WSN-based IoT monitoring, (3) smart grid billing services, etc., have been discussed in [32].

1.3 IoT and Cloud Computing Integration Architecture

The cloud and IoT contain many characteristics which are complementary to each other as shown in Table 1.1. So, their integration can provide good solutions to real-world problems. The IoT can gain from virtually unlimited storage resources be it storage or computing, and pervasive reachability of Cloud. The cloud can benefit from IoT by increasing its scope to address the real-world issues by providing new services in real life scenarios [11].

Table 1.1 Characteristics of IoT and Cloud

Cloud	IoT	Characteristics
Provide techniques to manage	Generator	Big data
Virtually unlimited	Limited or none	Storage capacity
Virtually unlimited	Limited	Computational capabilities
Centralized	Widespread	Displacement
pervasive	Limited	Reachability
Mode of service providing	Point of convergence	Internet role

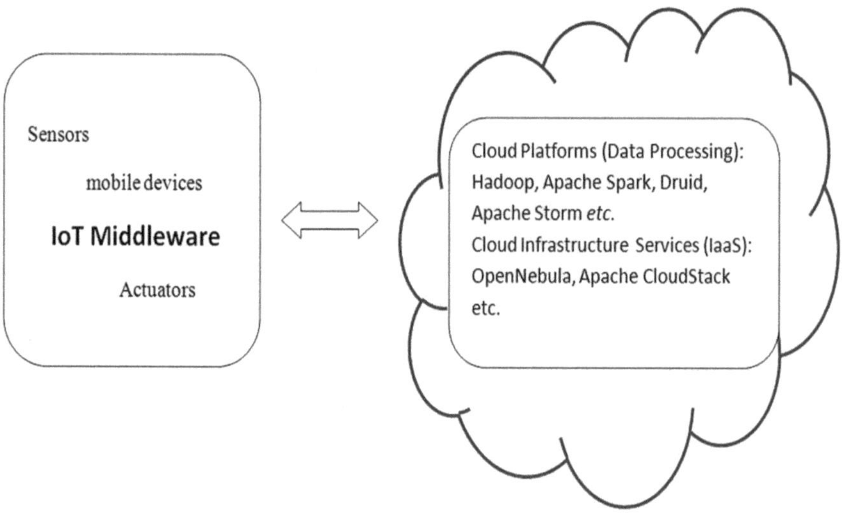

Fig. 1.3 Cloud and IoT integration paradigm

1.3.1 IoT and Cloud Integration Components

Integration components of cloud and IoT may be classified into three categories considering seamless integration that include cloud platforms and infrastructure, and IoT middleware. The cloud platforms address the IoT limitations and provide business opportunities and additional scalability. Cloud platforms are managed by the Cloud Infrastructures by deploying and monitoring. Middleware of the IoT provides a lightweight and interoperable procedure for exchanging data between the IoT devices and Cloud system deployed. Figure 1.3 shows the integration components [33].

1.3.1.1 Cloud Platforms

Since the IoT users are increasing every day, the data generated by them is also increasing, which has made the database management systems (DBMS) inappropriate. Therefore, a platform with high scalability, storage, and processing power is needed. We study various platforms for storing large amount of data as well as processing that data.

(a) *Hadoop*: Hadoop [34, 35] is a freely available software framework, which manages big data by distributed storage and processing using MapReduce programming model. It is divided into four modules: Hadoop Common, Hadoop Distributed File System (HDFS), MapReduce, and Hadoop YARN. The Hadoop MapReduce is an implementation of MapReduce programming model used for writing applications that can process large-scale data of order of volume of multi-terabyte datasets in parallel on huge clusters in terms of thousands of nodes of commodity hardware in a safe, fault-tolerant manner. The MapReduce framework divides the input dataset into autonomous chunks that are organized by the map tasks in a parallel mode. The framework is responsible for scheduling tasks, monitoring them, and re-executing the failed tasks. YARN is added in Hadoop 2.0 framework with MapReduce programming model for cluster resource management. The YARN borrowed the resource management functions from the MapReduce model by including a separate layer to manage the resources and allowing new programming models in parallel. Currently only the MapReduce application runs on top of it. Furthermore, the intricacy to develop programs in MapReduce has led to new systems that convert programs into MapReduce, like Apache Hive [54], a data warehouse like SQL for managing and querying volumetric data, and Apache Pig [55], a platform for analyzing large datasets with a high-level language for expressing programs.

(b) *Druid*: Druid is an open-source, column-oriented, distributed analytics data store designed for business intelligence OnLine Analytical Processing (OLAP) queries on data [36, 37]. The platforms storing huge amounts of information as Hadoop do not assure about the time in which the information can be accessed or stored and also query mechanism is not efficient. The Druid addresses these issues by providing a real-time data ingestion, fast data aggregation, and flexible data exploration, and is mostly used in user-facing analytics applications. The Druid architecture mainly consists of two parts: *historical nodes*, to store and query the non-real-time information, and *real-time nodes* that can consume the stream information and respond to queries related to this data. It has other components too: *coordination nodes* for coordination of data management and data distribution on historical nodes, *brokers' nodes* to receive queries and route those queries to historical and real-time nodes, and *indexer nodes* for ingesting the real-time and batch data into the system. The real-time nodes provide real-time information and build up portions for matured information that they forward to the historical nodes that in turn hold the information in profound storage and transfer it in memory at whatever point the coordination hub needs it. It additionally uses the Apache Zookeeper, and also Hadoop segment to synchronize the components in a cluster, to deal with the present cluster state, and the MySQL for keeping the metadata related to the information sections.

(c) *ApacheHbase*: It is an open-source, non-relational, and distributed database used for performing read and write of big data randomly or in real time [38]. ApacheHbase is developed as a part of Apache Software Foundation's Apache Hadoop project modeled after the Google's Bigtable, and written in JAVA. It works on the top of HDFS having potential of providing Bigtable like capabilities for Hadoop. The data storage in ApacheHbase is done like traditional relational database management systems (RDBMS) tabular method. ApacheHbase defines a four-dimensional information model with four coordinates and defines each cell. These four coordinates are (1) Row Key which is unique per row; it does not have a data type and is treated internally as a byte array, (2) Column Family which is unique per column and same for each row, (3) Column Qualifier defines actual columns, (4) Version is defines for each column and can have a configurable number of versions, and we can access the data for a specific version of a column qualifier. The HBase architecture is HMaster/Region Server in which the HMaster observes and manages all the Region Servers, and every Region Server serves observes and manages the underlying regions. In Hbase table, fault tolerance is provided and distributed by a basic component Aregion.

(d) *Apache Kafka*: Apache Kafka is an open-source platform for stream processing and distributed messaging created by Apache Software Foundation [39]. The goal of the project is to supply a high throughput, low latency platform for handling real-time data supply and is programmed in Scala and JAVA. Kafka is run as a cluster on one or more servers. Stream of records are stored in Kafka cluster as categories so-called *topics*. This Kafka cluster stores streams of records in classes called points. Each record comprises of a value, timestamp, and a key. Kafka has four main Application Programming Interfaces (API): (1) Producer API permits an application to distribute a surge of records to at least one Kafka subjects, (2) Consumer API permits an application to subscribe to at least one points and process the flood of records delivered to them, (3) Streams API permits an application to go about as a stream processor, expending an information stream from at least one themes and creating a yield stream to at least one yield themes, successfully changing the information streams to yield streams, (4) Connector API permits building and running reusable producers or consumers that interface Kafka themes to existing applications or information frameworks.

The Kafka is built with a publish/subscribe queue used for the streaming data. The message stream in Kafka constitutes the topics in which a producer publishes the messages and a consumer subscribes to receive them. Topics are divided for handling load balancing and fault tolerance; every division can have multiple replicas. Whenever a message is sent by a producer to Kafka, it gets saved in designated partition(s) in a disk and its copies are absorbed for fixed time duration. It allows group-consumers for load balancing ingestion information among the consumers. The producers–consumers can work synchronously or asynchronously with batch data; however, there is a trade-off between throughput and latency.

For processing cloud data in batch, distributed or real-time manner other platforms of cloud are Apache Storm, Spark Streaming, RabbitMQ [40], etc.

1.3.1.2 Cloud Infrastructure

Cloud computing primarily provides IaaS, PaaS, and SaaS services. In cloud data center IaaS is the key component that provides capabilities. The IaaS service model provides computing components with virtualization. The IaaS model provides virtualization for computing components like storage, network platforms. Several IaaS services are discussed below:

(a) *OpenNebula*: It is an open-source cloud computing infrastructure for management of virtualized, heterogeneous data centers to enable private, public, and hybrid clouds [41, 42]. OpenNebula main goal is to provide a flexible, open, extensible, and comprehensive management layer to automatize the process of enterprise clouds by leveraging and combining existing deployed solutions for networking, monitoring or user management, storage, and virtualization. In OpenNebula all components are grouped in single key component called cloud OS. Cloud OS used to govern virtual and physical infrastructures as well as manages the allocation of virtual resources. The OpenNebula architecture is based on a front-end/host type which contains a master node and each host works as a slave node. Mater node manages and monitors the cluster and slave nodes as a common cloud platform runs virtual machines.
(b) *IaaSOpenStack*: OpenStack project started in 2010, a joint work of Rackspace Hosting and NASA. It is a free and open-supply platform deployed as an IaaS and focusing on (private or public) cloud that has an AWS EC2 like Application Programming Interface (API) [43]. The OpenNebula has a centralized system with elective components, the OpenStack incorporates a set of interrelated sub-additives with its personal APIs, that are wanted to be pre-set up and integrated for OpenStack deployment. It has a pluggable peer structure that an aspect can be installed and controlled in a different node. It provides four main helps in terms of sprint-board, compute, networking, and storage. The dashboard manages OpenStack assets and services through a customizable internet-primarily based consumer interface (UI). The pc provider is a critical factor asset that manages and controls the cloud platform. The OpenStack has storage in the form of object and block. The object storage is for redundant, fault-tolerant, and scalable facts shops, whereas the block provides chronic block degree storage for performance sensitive eventualities. The networking carrier allows network-Connectivity-as-a-service for different OpenStack offerings that offers a user interface to realize the networks, supports many technology and networking companies through a pluggable architecture.
(c) *Apache CloudStack*: It is freely available software for infrastructure cloud services and used to set up public, private, and hybrid infrastructure cloud services [44]. CloudStack, freely available software, was developed to set up and manage broad virtual machines network, as a highly scalable and available IaaS cloud computing platform. CloudStack also has an AWS EC2 support for public IaaS cloud similar to the OpenNebula and OpenStack. Its architecture is also master/slave type in which there is a CloudStack Management Server that manages all the resources in cloud. It executes in an Apache Tomcat container and many hypervisor nodes employ the virtual machines in each node via installed hypervisor. The CloudStack control Server avails the internet user interface (UI) and application program interface (API), manages storage and images, and it can be set up in a multinode mode for excessive accessibility

and load balancing among the management servers. Just like OpenNebula, it helps zones to be able to control geographically allotted nodes. A dispensed geographical zone offers a higher stage of fault tolerance as a cloud device can regain from an exterior disaster in a region. The Cloud Stack storage is of two sorts: number one and secondary garage. The former shops the disks of all virtual machines handiest and later saves the disks, snapshots, ISO snap shots, and disk templates. It does no longer aid heterogeneous secondary garage; but, it has plugin facility to make the OpenStack object garage and Amazon S3 as secondary storage. It additionally affords VDC in zones, referred to as digital private Clouds (VPC), for separation facts middle deployment.

1.3.1.3 Middleware for IoT

This IoT layer may be considered as an abstraction of functionalities and communication capabilities of the devices in IoT deployment for pervasive integration with other technologies like cloud services. Below a description of different IoT middleware like GSN, DPWS, and LinkSmart is given.

(a) *Global Sensor Networks (GSN) Middleware*: GSN middleware project is started in late 2004 at EPFL (École polytechnique fédérale de Lausanne). The GSN system is created using the sensors that may be real or virtual sensors, connected together to make the required processing path. GSN project main goal is to develop a universal platform for sensor networks and distributed processing of information produced by wireless sensor networks [45]. GSN developed in JAVA and runs on one or more systems working as the backbone of acquisition network. In GSN data streams are managed through XML specification files and live data is ingested into the system via remote wrappers. Its architecture is container based consisting of two main layers: *virtual sensor manager (VSM)* and *query manager (QM)*. VSM is used to manage virtual sensors and their infrastructure and QM layer is used to parse, execute, and planning SQL queries. For configuring alerts, VSN has a configurable notification manager and a topmost interface layer to retrieve via web services. In VMS, the connections with devices are handled by wrappers that are available for Tiny OS and other devices.

    There is a big problem of data heterogeneity when we need to interpret and understand it, in spite of the fact that it supports the heterogeneous devices and accession level problems. XGSN [46] is an extension of GSN middleware; it deals with this problem by making observation and search tasks in an IoT. The XGSN provides the semantics data to virtual sensors through an expansion of the SSN ontology. XGSN regulates the annotation process about sensors, sensing devices, and their efficiency. Like GSN, the XGSN also has interfaces to manage the virtual sensors, query data, and also integration for storing and processing the stream data with the linked sensor middleware (LSM).

(b) *Device Profile for Web Services (DPWS)*: It is a large collection of WSs specifications for embedded as well as resource-constrained devices. The DPWS was developed for resource-constrained devices to make them capable of using secure web services and it is based on the service-oriented architecture (SOA). DPWS is based on many other web services specifications like WS-Addressing for advanced end point and message addressing, WS-Policy for policy exchange, WS-Security for managing security, WS-Transfer/WS-Metadata exchange for device and service description, WS-Discovery and SOAP-over-UDP for device discovery, WS-Eventing for managing subscriptions for event channels. The DPWS also partially established on W3Cs WSs standards: simple object access protocol (SOAP) 1.2, XML schema and WS-Addressing, Web Services Description Language (WSDL) 1.1; and specifies several other protocols for messaging, locating, security, and eventing. DPWS is implemented in many ways such as WS4D, implemented via C/C++ and many other via JAVA programming languages, is an open-source implementation [47]. These implementations are supported by various platforms. DPWS protocol stack is very large hence its integration with resource-constrained devices may be expensive. DPWS-compliant gateway can abstract the underlying IoT components while taking advantage of DPWS interoperability and semantics on embedded devices. LinkSmart® was initially created within the Hydra EU project for Networked Embedded Systems that can help developers to incorporate physical devices of various types into their applications through easy-to-use web services for managing any device.

(c) *LinkSmart*: It was initially designed within the Hydra EU project to build a middleware that can empower the developers to include physical devices of various types into their applications using convenient web services to manage any device. LinkSmart is based on a SOA [48]. It was designed by considering the commonly known problem of compatibility of heterogeneous devices and various protocols associated with devices.

1.3.2 CoAP and 6LowPan-Based Cloud and IoT Integration Architecture

For communication over internet between devices, the web servers (WSs) that include RESTFul and SOAP are used. SOAP WS works with exchange XML but in most of the WSs operates over HTTP; this is the main challenge for resource and energy limited devices. The CoAP, an application protocol for resource-constrained internet devices, e.g., WSN nodes and sensors, enables these devices to use the RESTful services with constrained capabilities. CoAP uses UDP rather than TCP, commonly used in HTTP, for lightweight communication between resource constraint devices. There are two main sublayers in CoAP architecture: request/response and messaging. The request/response sublayer provides communication and the messaging sublayer provides reliability and duplication of messages. CoAP provides GET, PUT, PUSH, DELETE messages requests to retrieve, create,

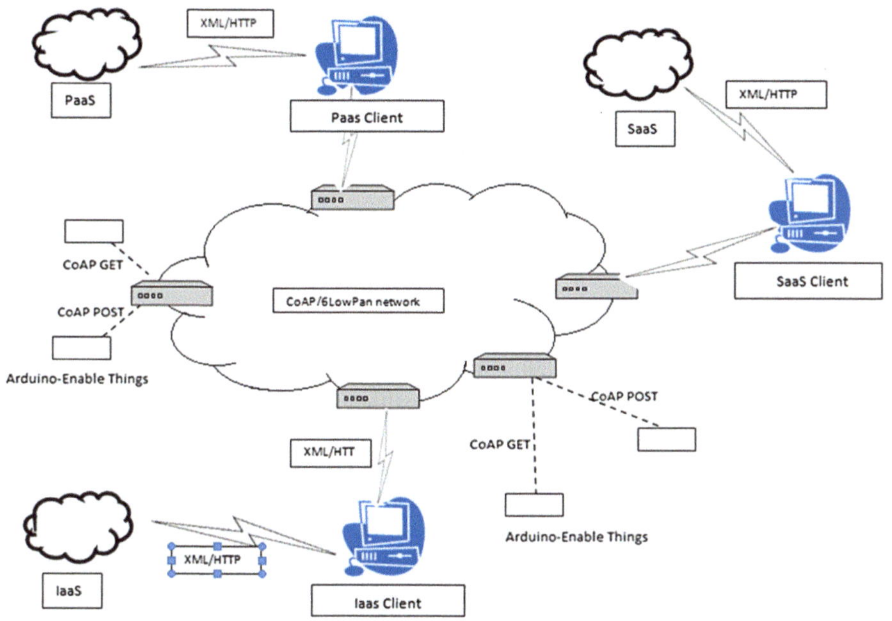

Fig. 1.4 Cloud IoT integration architecture based on CoAP and 6LowPan

update, and delete, respectively. The architecture of Cloud and IoT integration based on CoAP protocol between IoT devices, with RESTful services, and 6LoWPAN protocol for networking with things and their interaction with IaaS, PaaS, SaaS cloud computing models is shown in Fig. 1.4 [29].

1.4 Applications of Cloud Computing and IoT Integration

Integration of two rapidly growing technological areas makes sense for large number of applications [32], which is defined with characteristics, open issues, and challenges. Some applications of Cloud IoT are explained below.

1.4.1 Agriculture

IoT as a role: IoT in agriculture can be very useful as it can help in deciding more profitably crops with low cost in production [51]. The benefits of agriculture with IoT include monitoring the plants, soil, animals, controlling greenhouse environment, etc. The inputs used for agriculture like water, soil, pesticides, fertilizers, etc., along with their quantity and quality required for the crops are managed. *Cloud computing as a role*: In provincial areas, it is not viable for the farmers to manage administration suppliers on an individual premise. They require extensive and financial administration suppliers with various administrations. With IoT, a farmer has the capacity of crop delivery straightforward to the customers not just in a little local area, but in a more extensive region. This advancement will result in better crop and effective sales of the food product and production of food products beneficial to the real world.

1.4.2 Healthcare

The physiological data of a patient is maintained by some hospitals using the sensor networks [52]. The first contribution in healthcare field is IoT and multimedia services. To empower cost effective and high-quality ubiquitous medical services, efficient smart devices and cloud services are contributing for continuous and systematic innovation in healthcare. This area requires several applications such as hospital and physician networks, laboratories, health insurance pharmaceutical companies, patients, and other entities. The healthcare application generates a large amount of data (sensor data) that requires to be maintained in a proper manner for future processing and analysis. Mobile devices can make the services more efficient for delivery of health information in future for communication and access. The common challenges in this field are security, quality of service (QoS), interoperability, and dynamicity in storage.

1.4.3 Smart City

Typical middleware for future smart city can be given through IoT, obtaining information from sensing infrastructure, IoT technology (RFID sensors and geotagging), and putting data in a consistent manner, to strengthen the discovery, integration, and interconnection of actuators and sensors. This frames easy to applications for smart cities that are real-time and widespread connectivity [13]. This makes easier to third parties to develop IoT plugin to be connected to Cloud. The familiar issues are associated with security, real-time interaction, and resilience.

1.4.4 Smart Home and Smart Metering

To empower automation of regular in-home activities, the IoT has large applications for home atmosphere wherein the embedded devices have been used [53]. To build flexible applications with less code lines and to handle complex tasks, cloud is the best source to manage even huge data. When a single family smart home accessing reusable service is to be accessed online, some of the requirements should be satisfied: automation (home-based application should be attached to service provider like smart home-based Cloud), internal network inter connection (each intelligent digital thing in the home should interact with each other), intelligent remote control (smart home devices or objects can be managed or intelligently operated from anywhere). The existing literatures discuss metered solutions to provide identification of appliances, wireless sensor networks, and intelligent management of heating, lighting, consumption of energy, and air conditioning.

1.4.5 Video Surveillance

Video surveillance is one of the most important intelligent things as a part of security related issue because it works as a monitoring and self-management system [49]. The complex video analysis requires cloud-based solutions to fulfill the requirement of storage and processing. The video surveillance helps identifying, storing, and managing the video information from a camera and data delivery efficiently to a number of users through internet, load balancing, and fault-tolerance fashions.

1.5 Issues and Challenges for Integration

Some of the open issues in Cloud and IoT integration require further investigation that may be future directions [8, 33].

1.5.1 Security and Privacy Issues

Security and privacy are the main challenges to set up an IoT infrastructure. The devices in IoT are resource constrained, which can be exposed to attacks and threats. IoT devices sometimes use as well as generate sensitive information such as private information or vital infrastructure, which needs privacy to be included with devices, IoT network, and cloud infrastructure. The security and privacy aspects in applications of IoT in real world are discussed in [50]. To make IoT infrastructure secure several security techniques should be integrated with IoT. These security techniques involve protocol and network security for communication of heterogeneous devices of IoT, identity management for these heterogeneous devices, privacy for big data generated by IoT entities, trust and governance in IoT systems for communication and common framework, fault tolerance for attacks, and failure of IoT systems. The threats models, of an attacker, in the IoT are eavesdropping, node capture to gather data, denial of service (DoS), physical damage, and controlling various entities. Furthermore, Cloud computing acquisition in IoT also has security and privacy concerns.

1.5.2 Ipv6

One of the main components of IoT is internet which has address limitations due to IPv4 addressing scheme. The technologies for resource-constrained devices like CoAP can interact directly with embedded devices via Internet. Continuous growth of these technologies requires elimination of the network address translation (NAT) mechanisms so that they can address each IoT device or service with a unique IP address. The IPv6 addresses this problem using 128 bit IP address and also has several advantages like the increased range of devices connected to internet, source to destination connectivity, and an agreement with open REST interfaces [49]. In embedded devices of IoT, implementation of the IPv6 can be done by employing 6LoWPAN and ZigBee IP specifications, which still needs to be spread in many commercial platforms. The IoT network is trending towards networks of human-to-machine and machine-to-machine communications using IPv6 from networks with human initiated actions.

1.5.3 Need for Standards

In cloud and IoT paradigm, necessary protocols, architectures are required that is being standardized by the scientific community. It interconnects the enhanced services and heterogeneous smart objects to realize Cloud and IoT integration paradigm [23]. The important paradigm is Mobile-To-Mobile (M2M) with a little standard. So, the available solutions use internet, standard web, and cellular technologies. Most of the architectures at the primary phase of IoT are either from cloud or from wireless sensor networks.

1.5.4 Complex Data Mining

The issues related to big data cannot be addressed by the existing technologies. When high magnitude big data is generated, its high-level frequency, gap between the data availability, and its organization for processing get wider. Further research is required to address the challenges of big data, heterogeneous spatiotemporal (geo-related and sparsely distributed) data, i.e., the data that worth more mixed with erroneous data, are not directly consumable using virtualization platform in IoT. To create interesting and easy to perceive visualization, new methods are to be developed (e.g., 3D, geographic information system (GIS)).

1.5.5 Cloud Capabilities

In a networked environment, the security is one of the major issues for Cloud and IoT integration due to various attacks on both of IoT (i.e., RFID, WSN) and Cloud side. Integrity, confidentiality, and authenticity in IoT can be done using encryption that can address inside attacks and also implementable on processing or embedded devices. The RFID components achieve high-level security due to high-level intelligence. The QoS requirements of diverse users, seamless execution of applications, and domain specific programming tools are also required to deliver reliable services. Duplication for cloud scheduling algorithms can help in failure management.

1.5.6 Fog Computing

The next step to cloud computing is fog computing which is an intermediate between the edge of the network and Cloud to deal with latency-sensitive applications to meet their delay requirements [50]. Alike to cloud, the Fog services include computing, data storage, application services to target users. Future work in smart grid will develop Fog computing paradigm. Even Fog devices are being developed to interact directly to a Cloud.

1.5.7 Energy Efficiency

With the ubiquity of sensor networks and their availability with the cloud, this will unavoidably prompt a considerable measure of information correspondences, which devours a great deal of power. A commonplace wireless sensor hub is made out of four parts: sensing unit, processing unit, transceiver, and power unit. If there should be an occurrence of video sensing, video encoding what's more, translating, power assumes a key part. Typically, video encoding is more complex, as contrasted with interpreting. The purpose for this is for productive pressure, the encoder needs to dissect the repetition in the video. It won't be reasonable to have a transitory power supply, similar to batteries and need to supplant them from time to time. With billions of sensors and low power devices, it is past probability. Having effective use of energy and rather perpetual power supply would be required. There ought to be means for sensors to produce power from the condition, as sun oriented energy, vibration, and air. Moreover, successful rest mode can be exceptionally convenient in such manner also. Another arrangement exhibited in is bringing cloud assets privately, known as Fog computing. Mist alludes to a limited cloud, which can be utilized for process offloading reason for the fundamental IoT devices.

1.6 Conclusion and Future Scope

In cloud computing and IoT model, the IoT is both dynamic and universal networked infrastructure oriented, and manages self-configuring nodes (things) with high intelligence. Since the IoT has limited capabilities in terms of processing potential and storage, the issues related to performance, security, reliability, privacy are of major concern. The integration of IoT with Cloud Computing is more beneficial for the applications requiring unlimited capabilities such as storage and processing capability. An ample number of applications such as smart city, healthcare, smart home, smart metering, video surveillance, agriculture, and automotive can significantly be improved in CloudIoT or Cloud of Things paradigm. High-level architecture based on Cloud Assisted Computing manages the complexity of smart object-based Internet of Things. Since the cloud and IoT integration is in infancy stage, no standard architecture is available. We have presented major key challenges in CloudIoT system. Working on these challenges would contribute in standardizing the CloudIoT.

In order to standardize and to come up with the full potential of Cloud and IoT integration, further research efforts are required in multiple directions as mentioned.

- For identifying, naming, and addressing things in IoT, it's far required to guide huge range of IoT devices and mobility of them. Though the IPv6 can be an efficient answer, yet to accumulate it on a big scale is an ongoing work. Additional research on IPv6, for IoT infrastructure, is necessary to both speed up this slow process in specific scenarios (e.g. access networks) and to cope with new mobility and scalability requirements.
- For detecting environmental adjustments, the solutions primarily depend upon IoT information will permit the delivery of context-based totally offerings and offer the best offerings relying at the state of affairs. Such possibility will inspire the research for growing extra effective algorithms to supply the personalized contents and classified ads.
- Large scale aid for multi-networking including multihoming, connection handover, and roaming will be obligatory for enhancing the community reliability and making sure redundancy, continuous connectivity, QoS, and fault tolerance. In this context, the solutions based totally on software described networking are also envisaged.
- Many applications of Cloud and IoT integration will take advantage of efficient and adaptable mechanisms in designing logically isolated network parts over the global network infrastructures, which can be a crucial driving force for research in virtualization and software-described networking fields.

References

1. Devipriya S. Contribution of internet of things: a survey. J Web Develop Web Des. 2016;1(3):1–6.
2. Gubbi J, Buyya R, Marusic S, Palaniswami M. Internet of Things (IoT): a vision, architectural elements, and future directions. Futur Gener Comput Syst. 2013;29(7):1645–60.
3. Mell P, Grance T. The NIST definition of cloud computing. 2011.
4. Hamdaqa M, Livogiannis T, Tahvildari L. A reference model for developing cloud applications. In: CLOSER 2011—Proceeding of the 1st international conference on could computing and services science; 2011. p. 98–103.
5. Sheng Z, Wang H, Yin C, Hu X, Yang S, Leung VC. Lightweight management of resource-constrained sensor devices in internet of things. IEEE Internet Things J. 2015;2(5):402–11.
6. Aitken R, Chandra V, Myers J, Sandhu B, Shifren L, Yeric G. Device and technology implications of the internet of things. In: 2014 symposium on VLSI technology (VLSI-technology): digest of technical papers. IEEE; 2014. p. 1–4.
7. Sheng Z, Yang S, Yu Y, Vasilakos A, Mccann J, Leung K. A survey on the ietf protocol suite for the internet of things: standards, challenges, and opportunities. IEEE Wirel Commun. 2013;20(6):91–8.
8. Botta A, De Donato W, Persico V, Pescapé A. Integration of cloud computing and internet of things: a survey. Futur Gener Comput Syst. 2016;56:684–700.
9. Lee K, Murray D, Hughes D, Joosen W. Extending sensor networks into the cloud using amazon web services. In: 2010 IEEE international conference on networked embedded systems for enterprise applications (NESEA). IEEE; 2010. p. 1–7.
10. Botta A, De Donato W, Persico V, Pescapé A. On the integration of cloud computing and internet of things. In: 2014 international conference on future internet of things and cloud (FiCloud). IEEE; 2014. p. 23–30.
11. Alhakbani N, Hassan MM, Hossain MA, Alnuem M. A framework of adaptive interaction support in cloud-based internet of things (IoT) environment. In: international conference on internet and distributed computing systems. Springer International Publishing; 2014. p. 136–146.
12. Dash SK, Mohapatra S, Pattnaik PK. A survey on applications of wireless sensor network using cloud computing. Int J Comput Sci Eng Technol. 2010;1(4):50–5. E-ISSN: 2044-6004
13. Suciu G, Vulpe A, Halunga S, Fratu O, Todoran G, Suciu, V. Smart cities built on resilient cloud computing and secure internet of things. In: 2013 19th international conference on control systems and computer science (CSCS). IEEE; 2013. p. 513–518.
14. Rao BP, Saluia P, Sharma N, Mittal A, Sharma SV. Cloud computing for internet of things and sensing based applications. In: 2012 sixth international conference on sensing technology (ICST). IEEE; 2012. p. 374–380.
15. Zaslavsky A, Perera C, Georgakopoulos D. Sensing as a service and big data. 2013. preprint arXiv:1301.0159.
16. Prati A, Vezzani R, Fornaciari M, Cucchiara R. Intelligent video surveillance as a service. In: Intelligent multimedia surveillance. Berlin, Heidelberg: Springer; 2013. p. 1–16.
17. Guinard D, Floerkemeier C, Sarma S. Cloud computing, REST and mashups to simplify RFID application development and deployment. In: Proceedings of the second international workshop on web of things. ACM; 2011. p. 9.
18. Nimbits platform. http://www.nimbits.com. Accessed 10 Mar 2017.
19. Thingspeak platform. https://www.thingspeak.com. Accessed 10 Mar 2017.
20. Paraimpu. http://paraimpu.crs4.it. Accessed 10 Mar 2017.
21. Device cloud. http://www.idigi.com/devicecloud. Accessed 10 Mar 2017.
22. SensorCloud. http://www.sensorcloud.com.Accessed 10 Mar 2017.
23. Arduino. http://www.arduino.cc. Accessed 10 Mar 2017.
24. Stack4Things. http://stack4things.unime.it. Accessed 10 Mar 2017.
25. Wiring. http://wiring.org.co. Accessed 10 Mar 2017.

26. Sunspot. http://www.sunspotworld.com. Accessed 10 Mar 2017.
27. Mbed. http://mbed.org. Accessed 10 Mar 2017.
28. Hou L, Zhao S, Xiong X, Zheng K, Chatzimisios P, Hossain MS, Xiang W. Internet of things cloud: architecture and implementation. 2016. preprint arXiv:1609.07712.
29. Zhou J, Leppanen T, Harjula E, Ylianttila M, Ojala T, Yu C, Yang LT. Cloudthings: a common architecture for integrating the internet of things with cloud computing. In: 2013 IEEE 17th international conference on computer supported cooperative work in design (CSCWD). IEEE; 2013. p. 651–657.
30. Zhou J, Leppanen T, Harjula E, Ylianttila M, Ojala T, Yu C, Yang LT. Cloudthings: a common architecture for integrating the internet of things with cloud computing. In: 2013 IEEE 17th international conference on computer supported cooperative work in design (CSCWD). IEEE; 2013. p. 651–657.
31. Kumarage H, Ibrahim K, Abdulatif A, Zahir T, Xun Y. Secure data analytics for cloud-integrated internet of things applications. IEEE Cloud Comput. 2016;3(2):46–56.
32. Addo ID, Ahamed SI, Yau SS, Buduru A. Reference architectures for privacy preservation in cloud-based IoT applications. IJSC. 2014;2(4)
33. Díaz M, Martín C, Rubio B. State-of-the-art, challenges, and open issues in the integration of Internet of things and cloud computing. J Netw Comput Appl. 2016;67:99–117.
34. Shvachko K, Kuang H, Radia S, Chansler R. The hadoop distributed file system. In: 2010 IEEE 26th symposium on mass storage systems and technologies (MSST). IEEE; 2010. p. 1–10.
35. Hadoop, http://hadoop.apache.org. Accessed 10 Mar 2017.
36. Yang F, Tschetter E, Léauté X, Ray N, Merlino G, Ganguli D. Druid: a real-time analytical data store. In: Proceedings of the 2014 ACM SIGMOD international conference on management of data. ACM; 2014. p. 157–168.
37. Druid, http://druid.io. Accessed 10 Mar 2017.
38. Apache HBase, http://hbase.apache.org. Accessed 10 Mar 2017.
39. Apache Kafka, http://kafka.apache.org. Accessed 10 Mar 2017.
40. RabbitMQ, https://www.rabbitmq.com. Accessed 10 Mar 2017.
41. Moreno-Vozmediano R, Montero RS, Llorente IM. Iaas cloud architecture: from virtualized datacenters to federated cloud infrastructures. Computer. 2012;45(12):65–72.
42. OpenNebula, http://opennebula.org. Accessed 10 Mar 2017.
43. IaaSOpenStack, https://www.openstack.org. Accessed 10 Mar 2017.
44. Apache CloudStack, http://cloudstack.apache.org. Accessed 10 Mar 2017.
45. GSN, https://github.com/LSIR/gsn/wiki. Accessed 10 Mar 2017.
46. Calbimonte JP, Sarni S, Eberle J, Aberer K. XGSN: an open-source semantic sensing middleware for the web of things. In: TC/SSN@ ISWC; 2014. p. 51–66.
47. WSD4, http://ws4d.org. Accessed 10 Mar 2017.
48. LinkSmart, https://linksmart.eu. Accessed 10 Mar 2017.
49. Ziegler S, Crettaz C, Thomas I. IPv6 as a global addressing scheme and integrator for the Internet of Things and the cloud. In: 2014 28th international conference on advanced information networking and applications workshops (WAINA). IEEE; 2014. p. 797–802.
50. Roman R, Zhou J, Lopez J. On the features and challenges of security and privacy in distributed internet of things. Comput Netw. 2013;57(10):2266–79.
51. Biduaa KR, Patela CN. Internet of things and cloud computing for agriculture in India. International Journal of Innovative and Emerging Research in Engineering. 2015;2(12): 2394–3343.
52. Doukas C, Maglogiannis I. Bringing IoT and cloud computing towards pervasive healthcare. In: Sixth IEEE international conference on innovative mobile and internet services in ubiquitous computing (IMIS), Palermo, Italy, July 4–6 2012;2012. p. 922–926.
53. Soliman M, Abiodun T, Hamouda T, Zhou J, Lung CH. Smart home: Integrating internet of things with web services and cloud computing. In: 5th IEEE international conference on cloud computing technology and science (CloudCom), December 2013, vol. 2;2013. p. 317–320.
54. Apache Hive. 2017. https://hive.apache.org. Accessed 10 Mar 2017.
55. Apache Pig. 2017. https://pig.apache.org. Accessed 10 Mar 2017.

Chapter 2
A Self-Governing and Decentralized Network of Smart Objects to Share Electrical Power Autonomously

Amrutha Muralidharan, Horia A. Maior, and Shrisha Rao

Abbreviations

DPD	Decentralized power distribution
HPF	Highest power demand first
IoT	Internet of things
LPF	Least power demand first
SPOF	Single point of failure

2.1 Background

Much effort has been invested in the analyses of methods for allocating and managing resources among large distributed systems. Together with these issues, improving the energy efficiency of a system can be done at the demand-management level [21]. A common resource to be shared is electrical power, used in all kinds of devices around us, such as household appliances (fridge, computer, kitchen appliances, etc.). An Internet of Things (IoT) [9] is a term to describe a network of *things*—objects not traditionally thought of as computers (e.g., cars, household appliances), which may nonetheless be connected using Internet protocols and

A. Muralidharan
Microsoft India (R&D) Private Limited, Hyderabad, India
e-mail: amrutha.m@iiitb.org

H.A. Maior
School of Computer Science, University of Nottingham, Nottingham, UK
e-mail: maiorh@acm.org

S. Rao (✉)
International Institute of Information Technology - Bangalore, Bangalore, India
e-mail: shrao@ieee.org

W. Rivera (ed.), *Sustainable Cloud and Energy Services*,
DOI 10.1007/978-3-319-62238-5_2

technologies (TCP/IP, etc.). Assuming that such a network of objects is connected to a shared, variable power supply, it is fruitful to consider protocols by which they may function effectively in a decentralized manner, efficiently sharing electrical power. We present a novel model where a network of connected, autonomous smart *objects* efficiently share electrical power from a common, limited power supply.

Every object within the network has an individual power requirement (rating). The power rating of an object may change over time, as it may have multiple levels of power settings (e.g., a hair dryer or washing machine typically has multiple settings), and objects can either be powered (i.e., consuming their full power demand) or not powered at all (consuming zero power). Following Rao [25], objects drawing power resource are distinguished from one another by a priority value. This is because in practice, some objects are more important than others in terms of their functions and utilities. In our model, we assume that the priorities of objects in the network are provided to us, thereby enabling us to distinguish which objects should have access to power first. Kato et al. [13] propose "a dynamic priority model to control the power flow to the appliance in real time… because the priority of an individual appliance should be determined dynamically depending on the appliance properties and user's lifestyle pattern." Our work is agnostic to the method used to arrive at object priorities. The priorities may change over time (for example, a toaster may have a higher priority in the mornings than during the rest of the day).

The model we present in this paper overcomes the limitations of having a central controller or a master node of some sort to oversee the activities of all objects (making decisions on behalf of the objects [20]). Instead, all objects decide their own actions based on what we may consider an individual overview of the whole system. The overview would be the information collected by each object in the system from all other objects. This is possible through direct object communication (objects exchanging information among them without human intervention [24]): each object shares with other objects its priority and power demand by broadcasting messages over the network. It is also assumed, as is standard, that the power required for the objects to compute their actions is negligible compared to the power required for their main functions.

The resulting system is an extensible and decentralized one, also overcoming limitations such as scalability and reliability; objects can join or leave the system at any time. Such a system is also not subject to a *single point of failure* (SPOF). In other words, if an object fails to complete an action/task, the rest of the system would not stop working or would not be affected, unlike a centralized system which would be rendered entirely dysfunctional if the central controller were to fail. The decentralized approach requires every object in the system to have processing and communication functions, which would give them computing capabilities, and make them capable of decision making. Such self-governing objects are technologically feasible [23]. Liu and Zhou [14] describe an "autonomy feature," where objects have the ability to reason, negotiate, understand, adapt, and learn from other objects.

The power source for the system might not provide enough power to satisfy all objects in the network at once. Therefore, higher priority objects should have the chance to get powered first. To satisfy as many objects as possible given a limited power budget, lower-consuming high priority objects (higher-priority objects with smaller power demands) would be considered for powering first. We refer to our approach of power distribution in the network as *decentralized power distribution (DPD)*. Objects in the network might not necessarily require power at all times. For example, a toaster may be considered to need power in the mornings when toast is desired for breakfast, but it may not require power during the rest of the day. Such objects can exit the network, thereby ensuring that the local overview of the objects in the network contains only relevant information.

Our model follows a non-preemptive power allocation policy, with no partial fulfillment. When an object is allocated power, the allocation is not forcefully interrupted until it has finished its work, and allocation is either zero or sufficient to meet its demand. However, there could be objects like laptops (when there is sufficient charge), refrigerators, etc. which have "temporal interruption capability" [13]. When sufficient power is not available for allocation to a higher priority object, powered objects with temporal interruption capability can sacrifice their power for higher priority objects. Moreover, in the network there are no dependencies among objects; objects do not have to wait for other objects to finish tasks/jobs so they can do their own work.

In subsequent sections, we describe the model in detail; also how objects communicate with other objects, how they explore the system and create a local overview of the network, and how they inform other objects when they wish to exit the network. We also describe the Power Allocation and Redistribution algorithms in the network using DPD; how and in what order the objects get powered, and how the available power is distributed among the objects. Whether power is allocated to an object or not is decided only after running both Power Allocation and Redistribution algorithms, in that order. The Power Allocation algorithm considers objects in decreasing order of priorities and at a particular priority level, in increasing order of power demands. Thus the highest priority object with the smallest power demand is the first to be addressed by this algorithm. In the Power Redistribution algorithm, the power which was earlier assigned to a lower-consuming object (by the Power Allocation algorithm) could be reassigned to a higher consuming object at the same priority level in order to reduce the unallocated power in the system.

DPD ensures that the available power is used to the maximum extent possible. When it comes to sustainable energy, the available power resource is variable, and it should be used to the maximum extent possible at a given time (considering the example of solar power, as much of the available power as possible should be used when the sun is shining, as storage and later use tend to be expensive and problematic [16]).

To complete the development of our model of network, we prove the correctness of our approach in Sect. 2.4. We formulate four theorems stating relevant correctness properties, and give corresponding proofs. In Sect. 2.5, a simulation example is presented, showing for a sample network consisting of a set of household appliances and a given power budget, how and in which order, appliances get powered. Section 2.5 also compares the performance of DPD with two other methods of power distribution; least power demand first (LPF) and highest power demand first (HPF).

An IoT of smart objects has the potential to be a great method for solving demand-management problems [31]. Higgins et al. [8] discuss "a pathway" to flexible power system automation. They propose a new approach based on "distributed intelligence" rather than "traditional centralized control," with the system improving on many levels. We further develop along the lines of their approach, by creating a decentralized distributed model of an IoT, where power resource consumers can freely join and leave the system automatically at any time [28].

Zorzi et al. [32] point out that there is presently a first-time opportunity to use the IoT approach to interact with surrounding environments and to exchange information that previously was not available. They address a series of issues pertinent to the IoT paradigm, including connectivity, scalability, self-management, capability—all within the context of energy management. Monnier [17] discusses how objects around us need to be more and more connected and suggest that "connectivity is the key to automation." Perera et al. [23] survey context-aware computing by an IoT.

Niyato et al. [19] present a system that uses machine-to-machine (M2M) communication to reduce the costs of a home energy management system. They describe a smart grid system as having three major parts: power generation, power distribution, and power consumption. Their M2M communication takes place between home appliances and smart meters, and the system is dependent on a central node or control center [19]. DPD overcomes the limitation inherent in using a central controller of any sort. Objects in our model use a sort of M2M communication, but in a different context; objects communicate among themselves and decide their own actions (e.g., to get powered if suitable conditions exist). Karnouskos [12] supports such an approach, and points out that communication between objects and other entities, with "alternative energy resources which are smaller and decentralized," is one way to "achieve common goals such as energy efficiency." Walczak et al. [30] also describe M2M communication as part of IoT and Future Internet Engineering.

The combination of intelligent consumers and providers of sustainable energy has been recently combined conceptually as the "Internet of Energy" [1, 5]. Bui et al. [5] present a set of advantages of combining the smart grid and IoT, such as decentralization of the control procedures. They mention that from a communication perspective, the smart grid distribution network must also be scalable; this also appears necessary given the high number of devices that are presumed to take part in the future Internet of Energy [29].

Siano [26] presents ways in which systems can lower peak demand, and Gelazanskas and Gamage [7] propose a novel electricity demand control technique using real-time pricing. Such methods often require coercive stopping of consumption at a system-wide level. Our work overcomes the limitations of arbitrarily stopping objects (which are consumers attached to a smart grid) from consuming power at peak times, instead making the objects responsible for their own usage times (objects decide for themselves a suitable time to get powered) but subject to global availability limits.

The works discussed above consider how to reduce energy consumption in networked systems, but there is also a substantial literature (though not directly pertinent to our problem domain) on energy efficiency in networking itself—as may be seen in the survey by Bolla et al. [4].

2.2 System Description

In this section, an abstract model of the network within a demand-management problem is discussed. The definitions and notations of the problem are defined, along with a description of how the network is maintained.

Every object in the network is able to exchange information with other objects (for simplicity, we abstract away the details of how they are physically connected, and assume a reliable underlying communication service). It is also assumed that every object has some sort of reliable computing capability. When an object wishes to get powered, it joins the network and explores the whole system; it finds other objects in the network and stores information about them. This way, every object creates its own local overview of the system, and knows things such as its own priority relative to other objects and its own power demand relative to other objects at the same priority level as itself. Similarly when an object does not wish to get powered any more, it exits the network and informs all the other objects about its intention. The exploration and exit is possible through communication and exchange of information among the objects.

2.2.1 Notation

Let there be numerous objects ϱ_i that demand and may consume electrical power at any instant of time. The set of all such objects, denoted by O, together with the power supply and the underlying communication network, comprise the whole network at that instant $O = \{\varrho_0, \varrho_1, \ldots, \varrho_{n-1}\}$, the set of all n objects that demand electrical power from the power supply at time t.

Each object consumes a non-negative amount of power (i.e., objects are not themselves power sources). This amount of power, called the power demand of the object, can vary with time. Let T denote the set of all time instances. Let $\mathbb{R}_+$ be the set of positive real numbers, and $\mathbb{R}_+ \cup \{0\}$ the set of non-negative real numbers; then the demand function $f : O, T \rightarrow \mathbb{R}_+ \cup \{0\}$, where $f(\varrho_i, t) = r$ means that the power demand of object ϱ_i at time t is r.

The total power demand of the system at any time t, according to Rao [25], is the sum of power demands of all objects in the system at time t. The total power demand, denoted by $D : T \rightarrow \mathbb{R}_+ \cup \{0\}$, is given as:

$$D(t) = \sum_{i=0}^{n-1} f(\varrho_i, t)$$

The power supply can be described by the total amount of power from some source(s), available to be shared among all objects of the system at a given time. The power supply may also vary with time. The function $\gamma : T \rightarrow \mathbb{R}_+ \cup \{0\}$ is a power supply function, where $\gamma(t) = w$ indicates the power provided by the power source at a given time t.

Objects need to keep track of unallocated power in the system at any instant. This is the amount of power which is not allocated to any object and is hence available for allocation. The function $u : T \rightarrow \mathbb{R}_+ \cup \{0\}$ is the unallocated power supply function, where $u(\varrho_i, t) = v$ indicates that the power left unallocated in the system at a given time t is v. Objects also need to know if there was a reduction in the power supply. The function $r : O \times T \rightarrow \{\text{TRUE}, \text{FALSE}\}$ is used by each object to indicate that there was a reduction in power supply at time t. Just before the power distribution is done in the system or when there is a reduction in the power supply, $u(\varrho_i, t) = \gamma(t)$.

In Sect. 2.1 we described each object in the system as having an assigned priority level. Objects can have time-varying priority values. Let $\delta : O, T \rightarrow \mathbb{R}_+ \cup \{0\}$ be a priority function, where $\delta(\varrho_{i,t}) = a$ indicates that a is the priority of object ϱ_i at time t.

Objects need to keep track if they are getting powered or not. Considering the boolean values TRUE and FALSE, a status function $p : O \times T \rightarrow \{\text{TRUE}, \text{FALSE}\}$ is given by:

$$p(\varrho_i, t) = \begin{cases} \text{TRUE} & \text{if } \varrho_i \text{ is powered at } t \\ \text{FALSE} & \text{if } \varrho_i \text{ is not powered at } t. \end{cases}$$

When an object ϱ_i gets powered at time t, the boolean status function $p(\varrho_i, t)$, becomes TRUE. This way, each object knows its status (powered or not powered) at any given time.

There may be objects in the system which may be flexible to have preemptive power allocation at times, allowing better utilization of the available power. For example, under certain circumstances, power allocation to appliances like laptops, cell phones, etc. can be pre-empted to be allocated to higher priority objects. Let $d : O \times T \rightarrow \{\text{TRUE}, \text{FALSE}\}$ be a function that indicates if the power allocated to a given appliance may be pre-empted at a time t. This function is given by:

$$d(\varrho_i, t) = \begin{cases} \text{TRUE} & \text{if } \varrho_i \text{ can be pre-empted at } t \\ \text{FALSE} & \text{if } \varrho_i \text{ cannot be pre-empted at } t. \end{cases}$$

When joining the network, objects explore the system in order to create their local overview of the network. In particular, each object is interested in storing information about other objects; it wants to know the other priorities in the system *vis-à-vis* its own priority, and know of objects with the same priority as itself. Every object $\varrho_i \in \{O(t)\}$ maintains three arrays:

- $P_i[\cdot]$ is an array where object ϱ_i stores all the priorities in the system (but without the information of which objects are at what priority levels); this is sorted in decreasing order such that $P_i[0]$ holds the highest priority known to ϱ_i.
- $Q_i[\cdot]$ is an array denoting objects with the same priority as ϱ_i. $Q_i[\cdot]$ is sorted in increasing order of power demands, such that $Q_i[0]$ holds the id of the object with smallest power demand and same priority as ϱ_i.

– $R_i[\cdot]$ is an array denoting power demands of objects with the same priority as ϱ_i. $R_i[\cdot]$ is sorted in increasing order, such that $R_i[0]$ holds the power demand of the object with same priority as ϱ_i and highest power demand at that priority level.

When joining the system (just before exploring), every object ϱ_i has just one element in both $P_i[\cdot]$, $Q_i[\cdot]$ and $R_i[\cdot]$, *viz.*, ϱ_i itself. The arrays are continually modified as objects enter or exit the system.

2.2.2 Communication

As an abstraction of how the communication is done in the network, let $\varrho_i, \varrho_j \in \{O\}$ be two objects. Let $msg[\varrho_i, \varrho_j, \mu]$ be a message (when there is no confusion we simply write it as $msg[\cdot]$), where ϱ_i is the sender of the message, ϱ_j is the receiver of the message, and μ is the message (which may contain any kind of information). After receiving a message, the receiver ϱ_j stores the message, by function $store(\varrho_j, \mu)$, and sends an acknowledgement for receiving the message μ to the sender, as indicated by $ack[\varrho_j, \varrho_i, \mu]$. When an object ϱ_i sends a message μ to object ϱ_j, it locally sets a counter and waits for at most time C to get an acknowledgement from ϱ_j. If ϱ_i does not get an acknowledgement within this time, it assumes that there may be some sort of failure at ϱ_j and ϱ_j is no longer part of the system. ϱ_i broadcasts this message to all the objects in the system. T he objects can then update their local overview by deleting information about ϱ_j. Following a standard approach [22], a message can also be broadcast or "flooded" over a system of connected objects. Let $broadcast[\varrho_i, \mu]$ denote that a message μ is broadcast by an object ϱ_i.

2.2.3 Exploration

All objects which enter the network (objects previously not in the network but wish to get powered) need to "explore" the network. In this, objects communicate among themselves and exchange information about their priorities, power demands, and unallocated power in the system. At the same time, they receive, compute, and store information about other objects, creating their own local overview of the system. When the priority or power demand of an object changes, it has to broadcast the new and the old values to all other objects. The other objects can then update their local arrays $P[\cdot]$, $Q[\cdot]$ and $R[\cdot]$ appropriately.

2.2.4 Exit

Objects in the network may no longer wish to get powered, and may exit the system. The leaving object informs all other objects about its intention. These other objects can then update their local overview of the system. The message sent by the leaving object contains three important information parts:

- its priority level $\delta(\varrho_i)$;
- its power demand $f(\varrho_i)$; and,
- whether or not it is powered ($p(\varrho_i, t)$).

In case the leaving object ϱ_i has the same priority as an object ϱ_j receiving the message, the latter has to remove the leaving object from the arrays $Q_j[\cdot]$ and $R_j[\cdot]$. In case the leaving object ϱ_i is the last one of its priority, the whole priority level is removed from the array $P_j[\cdot]$. Finally, if the leaving object was powered, the power allocated to it becomes available for other objects. Each object updates the value of the unallocated power in the system $u(t)$.

2.2.5 Failures

We have not focused on handling faults and failures in the system, but this is intentional and does not affect the validity of our work. We briefly indicate why.

Communication Failures In the current context, communication failures are easily handled because of some properties of the domain. It is accepted that reliable communications are anyway essential in a cyber-physical system that share electrical power [27], as communication failures can be exploited by attackers. Real-time scheduling and control of a cyber-physical system of smart objects is thus possible [6]. Power-line communication and other such mature technologies also exist [3] so that objects that can receive power will always also be able to communicate, unlike in a classical disconnected asynchronous system where failures may go undetected [2, 15]. If an object loses its communication, it can simply have the fail-safe behavior of stopping usage of power. Thus, we do not explicitly consider communication failures in our model.

Object Failures Object failures can be handled using a classical heartbeat mechanism [10]. Each object has a next-in-line and a before-in-line object associated with it. For an object ϱ_i, the next-in-object ϱ_j is one that has the same priority as ϱ_i, but the smallest power demand greater than that of ϱ_i. If ϱ_i has the highest power demand at its priority level, the next-in-line objects are objects at the next lower priority level. Similarly, the before-in-line object ϱ_k of ϱ_i is one that has the same priority as ϱ_i, but the highest power demand smaller than that of ϱ_i, or an objects at the next higher priority level after ϱ_i.

Each object periodically sends a heartbeat to the next-in-line and before-in-line objects. Objects that do not acknowledge the receipt of such a heartbeat, or which do not send the heartbeat, are assumed to have left the system. A message is broadcast to all the objects in the system when an object discovers that another has left. This is again a standard approach seen in many systems, and is not our original work, so we do not belabor details.

2.3 Power Allocation and Redistribution Algorithms

We start this section by comparing the key properties of DPD with two other published approaches: the knapsack approach [18] and Energy on Demand (EoD) [13]. We then describe the DPD in detail; with the algorithms to allocate and redistribute the available power among the objects in the network. The basic subroutines used in the algorithms are described first, followed by the pseudocode and brief description of the algorithms.

2.3.1 Comparison of DPD with Other Approaches

The knapsack approach [18] formulates the power distribution problem as a single knapsack problem with each object associated with a power demand and a profit which contributes for user's "quality-of-life" (corresponding to priority level in our model) and the power source as the capacity of the knapsack (Table 2.1). The objective is to allocate power from the source by maximizing the "quality-of-life" of users. When there are many low "quality-of-life" objects in the system, the high "quality-of-life" objects may get totally ignored in this approach. Thus, in the knapsack approach, there is no distinction between an object that accounts for a high "quality-of-life" and an object that accounts for a low "quality-of-life."

We introduce a term "sum of priorities" which is similar to "quality-of-life." In DPD, we do not maximize the overall sum of priorities of the powered objects in the network. Instead, we maximize the sum of priorities at each priority level (thereby maximizing the number of objects powered at each priority level) starting

Table 2.1 Comparison of properties of DPD with EoD and Knapsack method

Property	DPD	EoD	Knapsack method
Scalable	Yes	Yes	No
Decentralized	Yes	No	No
SPOF	No	Yes	Yes
Priority inversion	No	No	Yes
Preemptive	No	Yes	Yes
Complexity	$\mathcal{O}(k \cdot n)$	$\mathcal{O}(n)$	$\mathcal{O}(W \cdot n)$

from the highest priority level to the lowest. The knapsack approach does not have an objective function to minimize the unallocated power in the system. Our model allocates power in such a manner that higher priority objects are considered for power allocation before lower priority objects; and also, at each priority level, power is allocated to objects in such a manner that unallocated power in the system is minimum. Thus DPD is a multi-objective multiple knapsack problem where each priority level is considered as a knapsack with capacity of the knapsack being the unallocated power after power allocation to higher priority levels. The objectives are to maximize the "sum of priorities" in each knapsack and to minimize the unallocated power in the system. The knapsack approach computes an optimal solution to the power distribution problem using a dynamic programming approach with a computational complexity of $\mathcal{O}(W \cdot n)$ for n objects when the power source is W units. This method thus depends on the power source and is worse for large values of W. On the other hand, the message complexity of our method is $\mathcal{O}(k \cdot n)$ when there are n objects and k different priority levels in the system. This is much smaller since $k \ll W$ (generally).

We have already mentioned that EoD computes the priorities of objects dynamically. It proposes a solution for power distribution involving a centralized server which runs the power distribution algorithm and informs the objects how much and how long the power is allocated to them. The centralized approach is automatically susceptible to SPOF. In EoD, when an object enters the system, if the unallocated power in the system is not sufficient to meet its power demand, power allocated to a lower priority object is interrupted and allocated to it.

2.3.2 Basic Subroutines

In the **AllocatePower()** subroutine (Algorithm 1), if an object ϱ_i is not currently getting powered ($p(\varrho_i, t) = \text{FALSE}$) or there was a reduction in the power supply ($r(\varrho_i, t) = \text{TRUE}$), it checks if the available power is sufficient to meet its power demand. If it is sufficient, power is temporarily allocated to object ϱ_i ($t(\varrho_i, t) = \text{TRUE}$) and the value of available power $u(t)$ is updated. Here we use a status function $t : O \times T \to \{\text{TRUE}, \text{FALSE}\}$ which indicates if power was temporarily allocated to ϱ_i. This is a temporary allocation, since there is a possibility that the power allocated in this step may get redistributed to another object at the same priority level with a higher power demand by the **RedistributePower()** subroutine (Algorithm 2). When there is a reduction in power supply ($r(\varrho_i, t) = \text{TRUE}$) and ϱ_i is currently getting powered ($p(\varrho_i, t) = \text{TRUE}$), if the new value of power supply is not sufficient for ϱ_i, the power allocated to ϱ_i is withdrawn.

The **RedistributePower()** subroutine (Algorithm 2) redistributes power from an object of lower power demand to an object of higher power demand at the same priority level, thereby minimizing the unallocated power in the system. In Algorithm 2, *requestQ* denotes the index of the lower power demand object in the local variable Q_i of object ϱ_i, and *currentQ* denotes its own index. At each priority

Algorithm 1: AllocatePower()

```
w ← u(ϱ_i, t)
if (p(ϱ_i, t) = FALSE or r(ϱ_i, t) = TRUE) and f(ϱ_i, t) ≤ w then
    t(ϱ_i, t) ← TRUE
    w ← w − f(ϱ_i, t)
    u(ϱ_i, t) ← w
end
else
    (p(ϱ_i, t) = FALSE
end
```

level, objects which are not currently getting powered ($p(\varrho_i, t) = \text{FALSE}$) and were not allocated power in Algorithm 1 ($t(\varrho_i, t) = \text{FALSE}$), check if there is an object with lower power demand that is willing to sacrifice the power allocated to it. If the total of the sacrificed power and the unallocated power in the system is sufficient to satisfy the power demand of the current object, power is redistributed from the lower power demand object to the current object (lines 4–16, Algorithm 2). If the lower power demand object is not willing to sacrifice the power allocated to it, the current object can inform all other objects at the same priority level that there is no object which is willing to sacrifice power, by assigning the variable *requestQ* to $\perp$, a distinguished value.

The **ChooseNextObjectOrPriority**() subroutine (Algorithm 3) describes the order in which Power Allocation and Redistribution takes place in the network. In this subroutine, *currentP* and *currentQ* denote the index of ϱ_i in local variables P_i and Q_i, respectively. If ϱ_i is the last object in case of **System Level Power Allocation** (described in Sect. 2.3.3), or the first object in case of **Priority Level Power Redistribution** (described in Sect. 2.3.4) at its priority level, then the next-in-line object is one at the highest priority that is lower than the priority of ϱ_i (*nextPriority*). Otherwise, the next-in-line object is one with the same priority level as ϱ_i but with a higher power demand in case of **System Level Power Allocation** or a lower power demand in case of **Priority Level Power Redistribution**.

2.3.3 *System Level Power Allocation*

As we have mentioned, some objects might be more important than others, therefore we ought to distinguish among them in terms of priority levels. In case of the power supply not being sufficient to satisfy all objects, higher priority objects are satisfied first. We also note that objects can have different power demands—one object may need more power than others. In case of objects with the same priority level, the system first serves the lower-consuming object. This ensures that the available power is allocated to maximum number of objects in order of their priorities.

Algorithm 2: RedistributePower()

```
w ← u(ϱ_i, t)
if p(ϱ_i, t) = FALSE and t(ϱ_i, t) = FALSE and requestQ < currentQ and
  requestQ ≠ ⊥ then
    v ← R_i[requestQ]
    if f(ϱ_i, t) ≤ w + v then
        requestObject ← Q_i[requestQ]
        Send request[ϱ_i, requestObject]
        if msg = YES then
            p(ϱ_i, t) ← TRUE
            requestQ ← requestQ + 1
            w ← w + v − f(ϱ_i, t)
            u(ϱ_i, t) ← w
        end
        else
            requestQ ← ⊥
        end
    end
end
```

Algorithm 4 describes how the available power is allocated to the objects in the system. It is run by all objects independently. It describes how the power allocation starts from the highest priority object with the smallest power demand, and how each object informs other objects about its actions and the remaining available power. The first part of the algorithm (lines 3–6) concerns the highest priority object with the lowest power demand. It is the first object with the chance to get powered. The power allocation is already described in Algorithm 1. When an object has the chance to get powered, it sends a message to the next-in-line object, which can be either of the same priority but with a higher power demand, or one of the next lower priority, informing it of the unallocated power available in the system. This is described in Algorithm 3. The second part of Algorithm 4 (lines 7–11) describes the behavior of objects upon receiving such a message. In the first place, objects check if the message was directed to them (line 8). If so, it is their chance to get powered. The lowest priority object with the highest power demand is the last one to receive the message. It broadcasts the value of unallocated power to all the objects in the system (lines 2–5 in Algorithm 3). This ensures that all objects are aware of how much unallocated power is available in the system at any point of time.

The worst case message complexity of Algorithm 4 is $\mathcal{O}(k \cdot n)$ (where k is the number of priority levels in the system).

Algorithm 4 ensures that power is allocated to the maximum number of objects in order of their priorities. However, the unallocated power in the system can be further minimized without any change in the number of objects getting powered,

Algorithm 3: ChooseNextObjectOrPriority()

```
w ← u(ϱ_i, t)
if [flag = 0 and sizeof(Q_i[·]) = currentQ + 1] or [flag = 1 and currentQ = 0] then
    if sizeof(P_i[·]) = currentP + 1 then
        broadcast[ϱ_i, w]
    end
    else
        nextPriority ← P_i[currentP + 1]
        requestQ ← 0
        broadcast[ϱ_i, nextPriority, w, requestQ]
    end
end
else
    if flag = 1 then
        nextObject ← Q_i[currentQ − 1]
    end
    else
        nextObject ← Q_i[currentQ + 1]
    end
    Send msg[ϱ_i, nextObject, w, requestQ]
end
```

Algorithm 4: System Level Power Allocation

```
flag ← 0
t(ϱ_i, t) ← FALSE
if δ(ϱ_i, t) = P_i[0] and Q_i[0] = ϱ_i then
    AllocatePower()
    ChooseNextObjectOrPriority(flag)
end
Upon Receiving msg[·]
if [ϱ_i = nextObject] or [δ(ϱ_i, t) = nextPriority and Q_i[0] = ϱ_i] then
    AllocatePower()
    ChooseNextObjectOrPriority(flag)
end
```

by redistributing power among objects at the same priority level using Algorithm 5 (described in Sect. 2.3.4). This ensures that the wastage of power is minimized. This is critical especially when renewable sources of energy like solar energy are used, since storing them is extremely difficult.

2.3.4 *Priority Level Power Redistribution*

In Algorithm 5, Power Redistribution starts from the highest priority object with the highest power demand (line 3). At each priority level, the network tries to redistribute power from an object with lower power demand to one with a higher power demand, using Algorithm 2. When an object gets a request from a higher power demand object to sacrifice its power, it checks if it was allocated power in Algorithm 4. If so, it sends a message expressing its willingness to sacrifice the power allocated to it (lines 13–20). Otherwise, it informs the higher power demand object that it does not have any power to be sacrificed. When an object gets a chance to redistribute power, it sends a message to the next-in-line object, which can be either of the same priority level but with a lower power demand, or one of the next lower priority (Algorithm 3), informing it of the unallocated power in the system and the next object which may be willing to sacrifice power (*requestQ*). The second part of the algorithm (lines 8–12) describes the behavior of objects upon receiving such a message.

Algorithm 5: Priority Level Power Redistribution

1 $flag \leftarrow 1$
2 $last \leftarrow sizeof(Q_i[\cdot])$
3 **if** $\delta(\varrho_i, t) = P_i[0]$ ***and*** $Q_i[last] = \varrho_i$ **then**
4     $requestQ \leftarrow 0$
5   **RedistributePower**()
6   **ChooseNextObjectOrPriority**(*flag*)
7 **end**
8 **Upon Receiving** $msg[\cdot]$
9 **if** $[\varrho_i = nextObj]$ ***or*** $[\delta(\varrho_i) = nextPriority$ ***and*** $Q_i[last] = \varrho_i]$ **then**
10   **RedistributePower**()
11   **ChooseNextObjectOrPriority**(*flag*)
12 **end**
13 **Upon Receiving** $request[\cdot]$
14 **if** $t(\varrho_i, t) =$ TRUE **then**
15     $t(\varrho_i, t) \leftarrow$ FALSE
16   **Send** $msg[\varrho_i, \varrho_j, YES]$
17 **end**
18 **else**
19   **Send** $msg[\varrho_i, \varrho_j, NO]$
20 **end**

The worst-case message complexity of Algorithm 5 is $\mathcal{O}(k \cdot n)$ (where k is the number of priority levels in the system).

Algorithms 4 and 5 are run periodically to refresh the system (e.g., once a day, or once an hour). Algorithms 4 and 5 are also run when there is a change in the power supply or when there are objects leaving the system. Whenever there is a change in power supply (increase or decrease), the power source broadcasts the new value to the objects. Power allocation and Redistribution will have to be redone for the entire system in that case. Some objects which were initially getting powered may no longer be powered (if there is a decrease in power supply) or more objects may start to get powered (if there is an increase in power supply).

Where there is a decrease in power supply, powered objects that are flexible for preemptive power allocation sacrifice their power and the algorithms are run with the total power as the sum of the power supply and the power sacrificed by the pre-empted objects. The pre-empted objects are not totally removed from the allocation list, but their power may get reallocated to higher priority objects. When new objects enter the system, the unallocated power is distributed among them without disturbing the objects that are already getting powered currently. If the available power is not sufficient for the new object, the power allocated to a lower priority object which is willing to pre-empt the power allocated to it $d(\varrho, t)$ = TRUE can be reallocated to the new object.

When objects that are getting powered leave the system, the power allocated to them becomes available for the remaining objects. Algorithms 4 and 5 are then run to distribute the unallocated power among the remaining objects. When the power demand of an object increases, it checks if the unallocated power is sufficient to satisfy the additional power demand. If so, it takes the additional power, otherwise it stops getting powered and the power allocated to it is freed for the other objects. When there is a decrease in an object's power demand, the difference in its power demand is available for the other objects.

2.4 Proofs of Correctness

We have already mentioned that our model assumes that the objects are connected by a reliable communication service and each object has processing and communication capabilities. These are very critical for any distributed system to work. We also assume that the priorities are provided to us. Each object knows its priority and power demand at any instant and is able to broadcast these to other objects. We now identify a set of desirable properties of the smart object model, and prove the correctness of the DPD design by showing that they hold.

We claim that the system we propose is *wastage free*. By this, we mean that there is no object $\varrho_i \in O$ in the system that is not powered, i.e., ($p(\varrho_i, t)$ = FALSE), when in fact there is enough power resource to meet ϱ_i's demand ($f(\varrho_i, t) \leq u(t)$).

This is indicated formally as follows.

Theorem 1

$$\nexists \varrho_i \in O,\ (p(\varrho_i, t) = \text{FALSE})\ \textit{and}\ (f(\varrho_i, t) \leq u(t)).$$

Proof By contradiction, assume $\exists \varrho_i \in O$ in the system such that $p(\varrho_i, t) = \text{FALSE}$ while there is sufficient power to satisfy its demands, $(f(\varrho_i, t) \leq u(t))$. There are two cases to consider:

- Object ϱ_i is the highest priority object with the smallest power demand (among objects of the same priority); if $p(\varrho_i, t) = \text{FALSE}$, following line 2 in Algorithm 1 $f(\varrho_i, t) \nleq u(t)$. Power demand $f(\varrho_i, t)$ cannot be both $\leq u(t)$ **and** $\nleq u(t)$. Hence a contradiction.
- Otherwise, the object must have received a message from a higher priority object; if $p(\varrho_i, t) = \text{FALSE}$, similar to the above case, $f(\varrho_i, t) \nleq u(t)$. Power demand $f(\varrho_i, t)$ cannot be both $\leq u(t)$ **and** $\nleq u(t)$. Hence also a contradiction.

□

We also claim that DPD ensures that the unallocated power in the system is minimum and hence all available power is used by the objects to the greatest extent possible. By this, we mean that there are no two objects at the same priority level, with the lower-consuming object getting powered while the higher consuming object is not, though it is possible to reallocate power from the lower-consuming object to the higher consuming one and reduce the unallocated power in the system. There are no objects $\varrho_i, \varrho_j \in O$ at the same priority level $(\delta(\varrho_i, t) = \delta(\varrho_j, t))$, with ϱ_i, an object with a higher power demand $(f(\varrho_i, t) > f(\varrho_j, t))$ not getting powered while ϱ_j gets powered $(p(\varrho_i, t) = \text{FALSE}$ and $t(\varrho_j, t) = \text{TRUE})$, though there is sufficient unallocated power in the system such that it is possible to reallocate power from ϱ_j to ϱ_i $(f(\varrho_i, t) \leq f(\varrho_j, t) + u(t))$ and reduce the total unallocated power in the system.

This is indicated formally as follows.

Theorem 2 *$\nexists \varrho_i, \varrho_j \in O$ such that $(\delta(\varrho_i) = \delta(\varrho_j))$* ***and*** *$(f(\varrho_i, t) > f(\varrho_j, t))$* ***and*** *$(p(\varrho_i, t) = \text{FALSE}$ and $t(\varrho_j, t) = \text{TRUE})$* ***and*** *$(f(\varrho_i, t) \leq f(\varrho_j, t) + u(t))$.*

Proof By contradiction, assume $\exists \varrho_i, \varrho_j \in O$ such that $\delta(\varrho_i) = \delta(\varrho_j)$ **and** $f(\varrho_i, t) > f(\varrho_j, t)$ **and** $p(\varrho_i, t) = \text{FALSE}$ and $t(\varrho_j, t) = \text{TRUE}$ while there is sufficient unallocated power to reallocate power from ϱ_j to ϱ_i $(f(\varrho_i, t) \leq f(\varrho_j, t) + u(t))$. There are two cases to consider:

- Object ϱ_i is the highest priority object with the highest power demand (among objects of the same priority). If $p(\varrho_i, t) = \text{FALSE}$ while there exists an object ϱ_j at the same priority level, with lower power demand than ϱ_i and $p(\varrho_j, t) = \text{TRUE}$, then following lines 4–16 of Algorithm 2, $f(\varrho_i, t) \nleq f(\varrho_j, t) + u(t)$ **or** ϱ_j replied with *NO* to the request from ϱ_i (lines 13–15 Algorithm 2).
  In the former case, $f(\varrho_i, t)$ cannot be both $\leq f(\varrho_j, t) + u(t)$ and $\nleq f(\varrho_j, t) + u(t)$ at the same time. Hence a contradiction.

In the latter case, $t(\varrho_j, t)$ = FALSE (lines 13–20 Algorithm 5). But $t(\varrho_j, t)$ cannot be both TRUE and FALSE at the same time. Hence also a contradiction.

– Otherwise, object ϱ_i must have received a message from a higher priority object or an object at the same priority level but with a higher power demand. If $p(\varrho_i, t)$ = FALSE, following lines 4–16 of Algorithm 2, $f(\varrho_i, t) \nleq f(\varrho_j, t) + u(t)$ **or** ϱ_j replied with *NO* to the request from ϱ_i.
  In the former case, $f(\varrho_i, t)$ cannot be both $\leq f(\varrho_j, t) + u(t)$ and $\nleq f(\varrho_j, t) + u(t)$ at the same time. Hence a contradiction.
  In the latter case, $t(\varrho_j, t)$ = FALSE (lines 13–20 Algorithm 5). But $t(\varrho_j, t)$ cannot be both TRUE and FALSE at the same time. Hence also a contradiction.

□

We claim that there is no *priority inversion* in DPD, meaning that higher priority objects get powered first, if possible (that is if there is enough power available to satisfy their demands), before the lower priority ones. This means that there are no objects $\varrho_i, \varrho_j \in O$, with $i \neq j$, such that object ϱ_j has a higher priority than object ϱ_i, the available power is more than the power demanded by any of the ϱ_i and ϱ_j ($u(t) \geq f(\varrho_i, t)$ and $u(t) \geq f(\varrho_j, t)$), but the lower priority object ϱ_i is powered ($p(\varrho_i, t)$ = TRUE) while the higher priority one ϱ_j is not ($p(\varrho_j, t)$ = FALSE).

This is indicated formally as follows.

Theorem 3 *$\nexists \varrho_i, \varrho_j \in O$ such that ($\delta(\varrho_i, t) < \delta(\varrho_j, t)$)* ***and*** *($u(t) \geq f(\varrho_i, t)$ and $u(t) \geq f(\varrho_j, t)$), ($p(\varrho_i, t)$ =* TRUE ***and*** *$p(\varrho_j, t)$ =* FALSE*).*

Proof By contradiction, assume $\exists \varrho_i, \varrho_j \in O$, such that priority of object $\delta(\varrho_i, t) < \delta(\varrho_j, t)$ and the power available can satisfy any of the two objects with power resource $u(t) \geq f(\varrho_i, t)$ and $u(t) \geq f(\varrho_j, t)$. We also assume that $p(\varrho_i, t)$ = TRUE **and** $p(\varrho_j, t)$ = FALSE after running Algorithms 4 and 5. There are two cases to consider:

– Object ϱ_j is the one with the smallest power demand object and the highest priority; following line 3 in Algorithm 4, object ϱ_j gets powered first (therefore, before ϱ_i) and $p(\varrho_j, t)$ = TRUE. Object ϱ_j cannot be in the same time both powered $p(\varrho_j, t)$ = TRUE and not powered $p(\varrho_j, t)$ = FALSE (the assumption made). Hence a contradiction.
– Object ϱ_j is not the highest priority with the smallest power demand object; objects get powered when receiving a message dedicated to them (lines 7–8 Algorithm 4). Messages flow from the higher priority object towards the lower priority ones (Algorithm 2). If $p(\varrho_i, t)$ = TRUE **and** $p(\varrho_j, t)$ = FALSE, the message must have first arrived to object ϱ_i. Therefore $\delta(\varrho_i, t) \geq \delta(\varrho_j, t)$. But the priority comparison cannot be both $\delta(\varrho_i, t) < \delta(\varrho_j, t)$ **and** $\delta(\varrho_i, t) \geq \delta(\varrho_j, t)$ at the same time. Hence too a contradiction. □

When there is an object failure, we assume that the heartbeat mechanism (Sect. 2.2.5) ensures that the next-in-line or the before-in-line objects get to know about it and broadcast the message to the other objects. We claim that the model we propose is *SPOF* free, meaning that failure of an object to perform an action/task does not stop or affect the rest of the system. This means that there are no objects $\varrho_i, \varrho_j \in O$ such that failure of ϱ_i leads to failure of ϱ_j.

This is indicated formally as follows:

Theorem 4 *At any time t, $\nexists \varrho_i, \varrho_j \in O$ such that ϱ_j fails when ϱ_i fails.*

Proof Assume that ϱ_i is the next-in-line object after an object $\varrho_k \in O$. Suppose object ϱ_i fails when it is its chance to get powered. Assume $\exists \varrho_j \in O$, such that ϱ_j fails when ϱ_i fails. There are two cases to consider:

- Object ϱ_j is the next-in-line object after ϱ_i. By contradiction, assume ϱ_j waits for the message from ϱ_i for its chance to get powered. Since ϱ_i has failed, it does not send the message to ϱ_j and ϱ_j does not get a chance to get powered. But, when ϱ_k does not get an acknowledgement from ϱ_i within time C after sending a message, it broadcasts that ϱ_i has failed, to all the objects. Thus all the objects in the system delete the information about ϱ_i. Then ϱ_j becomes the next-in-line object after ϱ_k and hence it gets a message from ϱ_k (Algorithm 2). But object ϱ_j cannot both get a chance and not get a chance at the same time. Hence a contradiction.
- Object ϱ_j is any other object other than the next-in-line object to ϱ_i. Suppose object ϱ_l is the next-in-line object after ϱ_i. We already proved above that ϱ_l is not affected by the failure of ϱ_i. Thus ϱ_l gets a chance to be powered. Then ϱ_j gets the chance to be powered before or after ϱ_l and is not affected by the failure of ϱ_i.

□

2.5 Results and Discussion

In this section, we analyze the results of our work using a simulation example and compare the performance of DPD in terms of "sum of priorities" and "unallocated power" with other decentralized power distribution methods.

2.5.1 Simulation

A sample system consisting of objects from a hypothetical smart home as consumers of power is considered with realistic data from a published source (see Table 2.2). Given limited proposed power budgets, we simulated which objects are powered and in which order in various scenarios (three cases). Each object has a power demand (usage setting), and a priority level (from 1 to 10) denoting its importance relative to the other objects. The power resource budget is an amount of power in kilowatts (kW) available to share among the objects.

Table 2.2 Electrical appliances with their power demands and priorities (source: [11])

Object name	Abbreviation	Power demand (kW)	Priority
Electric shower	ES	7	6
Dishwasher	Dw	1.5	1
Washing machine	WM	1.5	2
Tumble dryer	TD	3	2
Iron	Ir	1.6	2
Toaster	Ts	1	8
Oven	Ov	2	8
Grill	Gr	1.8	8
Fridge-freezer	FF	0.4	10
Vacuum cleaner	VC	1	3
Hairdryer	Hd	1	9
Plasma TV	TV	0.4	5
Desktop computer	PC	0.15	9
Broadband router	BR	0.01	10
Smart phone	SP	0.005	10

According to Algorithm 4, the highest priority object with the smallest power consumption is allocated power first. Algorithm 5 redistributes power from lower power demand objects to higher power demand objects at the same priority level whenever possible. Figure 2.1 shows that object SP should be powered first, and also describes the order in which power flows among the objects.

For a power availability of 23 kW, all objects are powered, as the total demand of all objects in the system is 22.365 kW < 23 kW. However, sometimes the budget may not satisfy the system needs, and at other times there may be objects that do not require power. We present three cases below, simulating the proposed system while varying power budget and object demands:

CASE 1: Simulating which objects get powered if the power budget is 5 kW and all objects in Table 2.2 require power and are hence in the system. Algorithm 4 allocates power to SP, BR, FF, PC, Hd, Ts, Gr, and TV. Their demand is 4.765 kW and therefore the unallocated power after running Algorithm 4 is 0.235 kW. Algorithm 5 redistributes power from Gr to Ov and the unallocated power thus reduces to 0.035 kW. Table 2.3 shows which objects are powered and which are not. Object ES is skipped as its demand is larger than the remaining power, and object TV is being powered. The remaining power 0.035 kW cannot satisfy any of the remaining objects or used for redistributing power, therefore this amount is not used and objects Gr, ES, VC, WM, Ir, TD, and Dw are not powered.

CASE 2: The power budget remains the same (5 kW), but objects Ov and Ts do not require power at the current time and hence leave the system. The second column of Table 2.3 shows how a lower priority object VC, that was not powered in Case 1 is powered in Case 2 for the same amount of power available.

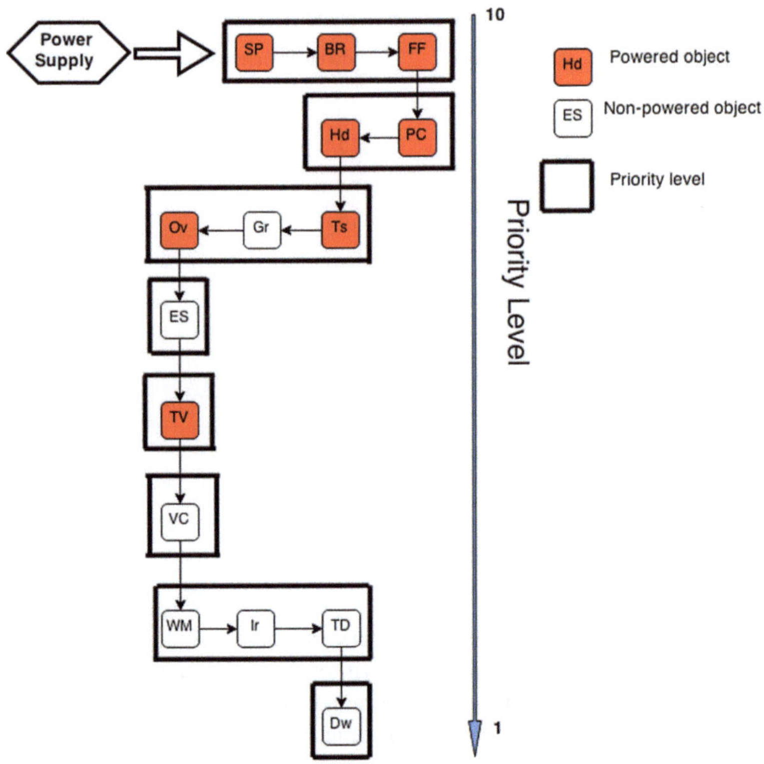

Fig. 2.1 The order of power usage in the IoT sample

CASE 3: Objects Ov and Ts still do not require power at the current time, and the power budget is increased to 11 kW. Table 2.3 shows how for a higher power budget, fewer objects are getting powered. This is because there is now enough power to satisfy object ES, and this object in particular has the highest power demand in the system (7 kW).

2.5.2 *Performance of DPD*

We already have noted that DPD ensures that power is allocated to the most number of objects in order of their priorities (from highest to lowest) and the unallocated power is minimized. In this section, we compare the values of "sum of priorities" and "unallocated power" obtained using DPD with two other methods; least power demand first (LPF) and highest power demand first (HPF). LPF allocates power starting from the object with the least power demand and then considering objects in increasing order of power demands. On the other hand, HPF allocates power to objects in decreasing order of power demand. When there are multiple objects with the same power demand, LPF and HPF select higher priority objects first.

Table 2.3 State of each object

Object	Case 1 (5 kW)	Case 2 (5 kW)	Case 3 (11 kW)
SP	On	On	On
BR	On	On	On
FF	On	On	On
PC	On	On	On
Hd	On	On	On
Ts	On	N/A	N/A
Gr	Off	On	On
Ov	On	N/A	N/A
ES	Off	Off	On
TV	On	On	On
VC	Off	On	Off
WM	Off	Off	Off
Ir	Off	Off	Off
TD	Off	Off	Off
Dw	Off	On	Off

On: object is powered; Off: object is not powered; N/A: object does not require power currently

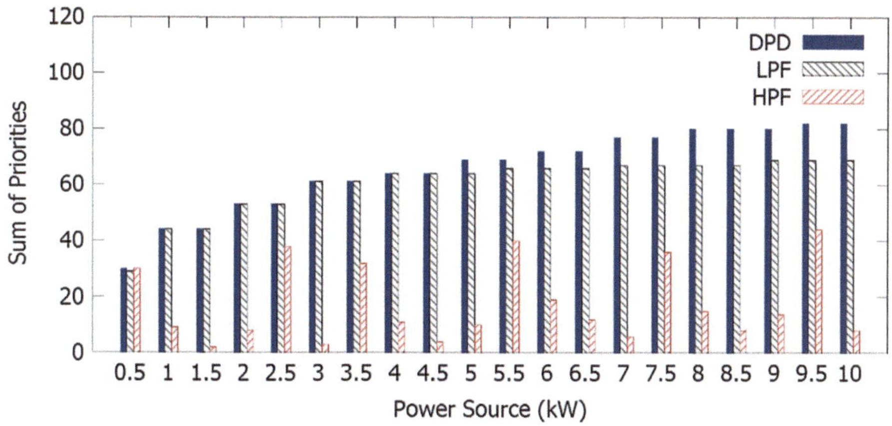

Fig. 2.2 Sum of priorities vs. power source in DPD, LPF, and HPF

In DPD, we do not attempt to simply maximize the "sum of priorities," since powering a large number of smaller priority objects may give the same value of "sum of priorities" as powering fewer number of higher priority objects. Figure 2.2 shows that DPD nonetheless may give the highest values for "sum of priorities." This is because it allocates power to higher priority objects first, and maximizes the number of objects selected at each priority level. In DPD and LPF, the objects which get powered at a lower power source value continue to get powered as the power available from the source increases and additional objects start getting powered. Hence the "sum of priorities" continuously increases when the power source output

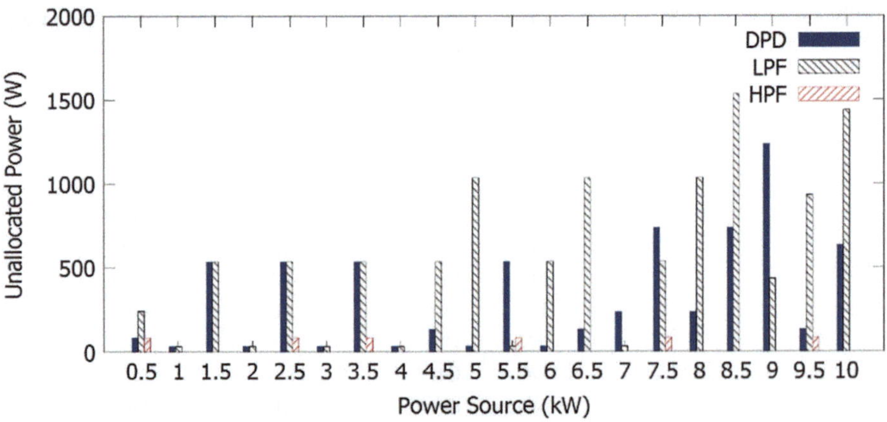

Fig. 2.3 Unallocated power vs. power source in DPD, LPF and HPF

increases for these two methods. On the other hand, in HPF, objects which were getting powered at lower power source need not continue to get powered at higher values. Hence the "sum of priorities" may increase or decrease with increase in power source. Figure 2.2 clearly shows that DPD selects the most number of higher priority objects compared to the other LPF and HPF. HPF selects lower priority objects for powering (indicated by the values of "sum of priorities") which is undesirable.

Figure 2.3 plots "unallocated power" obtained by DPD, LPF, and HPF over a range of values of power source. Clearly HPF has least unallocated power (less than 100 W mostly), and hence there is minimum power wastage when HPF is used. DPD performs better than LPF in terms of power wastage. We have already mentioned that DPD does not attempt to minimize the overall unallocated power in the system. It allocates power to objects such that as many as possible higher priority objects are powered and for a particular set of priorities chosen, the unallocated power is minimum.

2.6 Conclusions

We propose a model for the development of a self-governing system of smart objects that autonomously share power without a central controller. The DPD method describes an approach to distribute the available power among the appliances in the system in the best possible way based on their priorities. By "best possible," we mean that as many higher-priority objects as possible are powered with reduced power wastage. We do not look into how the priorities of objects are computed, and assume that they are provided to us. We also compare the properties and analyze the performance of DPD with other approaches and show that DPD is the best approach in a decentralized setup. Our model can be adopted and extended when implementing a real smart home or similar system that has to work with a smart grid or a variable power source. The model developed is extensible and decentralized, such that it meets the needs and requirements of a system with a variable number of connected objects. It is also SPOF free, and overcomes the limitations of having a central controller by having self-governing objects.

Some challenges in using this approach in practical systems would be as in other networked and IoT systems: to ensure system security, to be able to process data in real time, and to ensure network efficiency and reliability. The biggest difficulty could possibly be with ensuring compatibility of devices (objects) made by different manufacturers. This in turn would necessitate the development of common protocols and standards that would readily enable interconnectivity among devices. We see such development as a natural future outgrowth of our work.

References

1. Appelrath H-J, Terzidis O, Weinhardt C. Internet of energy — ICT as a key technology for the energy system of the future. Bus Inf Syst Eng. 2012;4(1):1–2.
2. Attiya H, Welch J. Distributed computing: fundamentals, simulations, and advanced topics. 2nd ed. New York: Wiley; 2004.
3. Berger LT, Schwager A, Pagani P, Schneider D. MIMO power line communications: narrow and broadband standards, EMC, and advanced processing. Boca Raton: CRC Press; 2014.
4. Bolla R, Bruschi R, Davoli F, Cucchietti F. Energy efficiency in the future internet: a survey of existing approaches and trends in energy-aware fixed network infrastructures. IEEE Commun Surv Tutorials. 2011;13(2):223–44 (Second Quarter 2011). doi:10.1109/SURV.2011.071410.00073.
5. Bui N, Castellani AP, Casari P, Zorzi M. The Internet of energy: a web-enabled smart grid system. IEEE Netw. 2012;26(4):39–45. doi:10.1109/MNET.2012.6246751.
6. Facchinetti T, Della Vedova ML. Real-time modeling for direct load control in cyber-physical power systems. IEEE Trans Ind Inf. (2011);7(4):689–98.
7. Gelazanskas L, Gamage KAA. Demand side management in smart grid: a review and proposals for future direction. Sustain Cities Soc. (Feb. 2014);11:22–30. doi:10.1016/j.scs.2013.11.001.
8. Higgins N, Vyatkin V, Nair N, Schwarz K. Distributed power system automation with IEC 61850, IEC 61499, and intelligent control. IEEE Trans. Syst. Man Cybern C. 2011;41(1): 81–92.
9. Holler J, Tsiatsis V, Mulligan C, Avesand S, Karnouskos S, Boyle D. From machine-to-machine to the Internet of things: introduction to a new age of intelligence. Amsterdam: Academic Press; 2014.
10. Hou Z, Huang Y, Zheng S, Dong X, Wang B. Design and implementation of heartbeat in multi-machine environment. In: 17th international conference on advanced information networking and applications (AINA 2003) (Mar 2003). p. 583–6.
11. How much electricity am I using? [Online]. Available: http://www.cse.org.uk/advice/advice-and-support/how-much-electricity-am-i-using.
12. Karnouskos S. The cooperative Internet of Things enabled smart grid. In: Proceedings of the 14th IEEE international symposium on consumer electronics, Braunschweig, Germany; 2010. p. 7–10.
13. Kato T, Yuasa K, Matsuyama T. Energy on demand: efficient and versatile energy control system for home energy management. In: Proceedings of the 2nd IEEE SmartGridComm. Piscataway: IEEE; 2011. p. 392–7.

14. Liu Y, Zhou G. Key technologies and applications of Internet of things. In: 2012 Fifth International Conference on Intelligent Computation Technology and Automation (ICICTA); 2012. p. 197–200.
15. Lynch NA. Distributed algorithms. San Francisco: Morgan Kaufmann; 1996.
16. Maity I, Rao S. Simulation and pricing mechanism analysis of a solar-powered electrical microgrid. IEEE Syst J. 2010;4(3):275–84. doi:10.1109/JSYST.2010.2059110.
17. Monnier O. A smarter grid with the Internet of Things. Texas Instruments White Paper (2013).
18. Morimoto N, Fujita Y, Yoshida M, Yoshimizu H, Takiyamada M, Akehi T, Tanaka M. Optimizing power allocation to electrical appliances with an algorithm for the knapsack problem. In: Proceedings of the 17th IEEE international symposium on consumer electronics. Piscataway: IEEE; 2013. p. 155–6.
19. Niyato D, Xiao L, Wang P. Machine-to-machine communications for home energy management system in smart grid. IEEE Commun Mag. 2011;49(4):53–9.
20. Palattella MR, Accettura N, Vilajosana X, Watteyne T, Grieco LA, Boggia G, Dohler M. Standardized protocol stack for the Internet of (important) Things. IEEE Commun Surv Tutorials. 2012;15(3):1389–406.
21. Palensky P, Dietrich D. Demand side management: demand response, intelligent energy systems, and smart loads. IEEE Trans Ind Inf. 2011;7(3):381–8.
22. Peleg D. Distributed computing: a locality-sensitive approach, vol. 5. Philadelphia: SIAM; 2000.
23. Perera C, Zaslavsky A, Christen P, Georgakopoulos D. Context aware computing for the Internet of Things: a survey. IEEE Commun Surv Tutorials. 2014;16(1):414–54. doi:10.1109/SURV.2013.042313.00197.
24. Prasad SS, Kumar C. An energy efficient and reliable internet of things. In: 2012 International conference on communication, information & computing technology (ICCICT). Piscataway: IEEE; 2012. p. 1–4.
25. Rao S. A foundation of demand side resource management in distributed systems. Trans Comput Sci. 2010;VIII:114–26. doi:10.1007/978-3-642-16236-7_8.
26. Siano P. Demand response and smart grids—a survey. Renew Sust Energ Rev. 2014;30:461–78.
27. Sridhar S, Hahn A, Govindarasu M. Cyber-physical system security for the electric power grid. Proc IEEE. 2012;100(1):210–24.
28. Suzdalenko A, Galkin I. Instantaneous, short-term and predictive long-term power balancing techniques in intelligent distribution grids. In: Camarinha-Matos LM, Tomic S, Graça P, editors. Technological innovation for the Internet of Things, vol. 394, IFIP Advances in Information and Communication Technology. Berlin/Heidelberg: Springer; 2013. p. 343–50.
29. Vermesan O, Friess P, Guillemin P, Gusmeroli S, Sundmaeker H, Bassi A, Jubert IS, Mazura M, Harrison M, Eisenhauer M, Doody P. Internet of Things strategic research roadmap. Internet of Things-global technological and societal trends. Aalborg: River Publishers; 2011. p. 9–52.
30. Walczak D, Wrzos M, Radziuk A, Lewandowski B, Mazurek C. Machine-to-machine communication and data processing approach in future Internet applications. In: 8th international symposium on communication systems, networks & digital signal processing (CSNDSP 2012). 2012. p. 1–5.
31. Wei C, Li Y. Design of energy consumption monitoring and energy-saving management system of intelligent building based on the Internet of Things. In: International conference on electronics, communications and control (ICECC 2011). 2011. p. 3650–52. doi:10.1109/ICECC.2011.6066758.
32. Zorzi M, Gluhak A, Lange S, Bassi A. From today's Intranet of things to a future Internet of Things: a wireless-and mobility-related view. IEEE Trans Wirel Commun. 2010;17(6):44–51.

Chapter 3
Implementing Energy Service Automation Using Cloud Technologies and Public Communications Networks

Claudia Battistelli, Padraic McKeever, Stephan Gross, Ferdinanda Ponci, and Antonello Monti

3.1 Energy System Automation Digitalization

3.1.1 What Does "Energy System Automation" Mean?

Power systems have been undergoing a process of automation since the second half of the last century. This process evolved with implementation in electromechanical technology over a period of 20 years, and then, since the 1970s and over the following three decades, with a progressive digitalization that has regarded protection, monitoring, control, and operation. Today, the automation process can be considered complete in the HV sections of the network, limited to primary substations in the MV sections, and not substantially present in the LV sections. From the 1990s onwards, automation has been gradually introduced into MV/LV distribution—also under the drive of the new market mechanisms of energy deregulation—and progressively increased, albeit slowly due to the complexity and inhomogeneous configuration of the distribution grid.

Today, automation is strongly boosted by the development of Information and Communications Technology (ICT), in connection with the new paradigms, methodologies, and technologies of Smart Grid (SG) and smart Microgrids (MGs). By now, it is a fact that, since early 2000, every aspect related to power systems, at any voltage level and network layer, has been increasingly developed in the prospects of the SG.

In this context, the future trends for SG will be increasingly characterized by ICT application to the existing grid, with the aim of making it smart. As the history of the power system until today tell us, this means that the whole network—up to

C. Battistelli, Ph.D. (✉) • P. McKeever, M.Sc • S. Gross, M.Sc • F. Ponci, Ph.D. • A. Monti, Ph.D.
Institute for Automation of Complex Power Systems (ACS), RWTH Aachen, Aachen, Germany
e-mail: CBattistelli@eonerc.rwth-aachen.de

W. Rivera (ed.), *Sustainable Cloud and Energy Services*,
DOI 10.1007/978-3-319-62238-5_3

the LV sections—will become completely automated in the years to come, with digitization enabled by the most modern ICT, and applied in the smarter and fully integrated available versions.

3.1.2 Introductory Concepts

ICT has experienced a period of rapid technological development in the last few years. Industry-standard ICT solutions are broadly applicable across many different industrial fields, and so the SG domain can also benefit from new ICT developments, saving development time and cost compared to the traditional method of developing industry-specific once-off fragmented solutions to each individual problem.

The development of SG is currently being held back by fragmentation: the industry is characterized by many different actors (whole roles vary according to the legal framework of the country of operation), by geographical diversity (where companies are generally local in scope), by the significant footprint of incumbent utilities (whose corporate cultures lie embedded in traditional grids), and by lack of open information flow between actors. Furthermore, many power and energy companies are not strong in the area of modern ICT.

3.2 Cloud Infrastructure, from Hardware to Software: General Concepts and State of the Art

The National Institute of Standards and Technology (NIST) provides a widely accepted definition of Cloud Computing [1]. It defines *Cloud Computing* as a shared pool of freely allocatable computational resource that can be managed with a minimum amount of service provider interactions. Today, many cloud service providers are available, like Amazon's AWS program, Microsoft's Azure, and Google's Cloud Platform. They offer a large variety of service from provision of storage, CPU, and network resources to highly advanced machine learning algorithms for data clustering or supervised learning. This large selection of services can be overwhelming at the beginning. That is why this chapter gives an overview of the most important developments in Cloud Computing in the last years and a basic setup for vendor independent cloud applications.

Cloud Computing utilization increased massively in the last years, initiated by the start of Amazon's Elastic Compute Cloud service in 2006. At the beginning of this period, early adopters needed to develop their applications specifically for a vendor's infrastructure. This resulted in various highly specialized and mostly proprietary software platforms and applications. Nowadays, there are various open source tools and technologies available that simplify the development of cloud applications. Especially the arising of a series of general purpose cloud

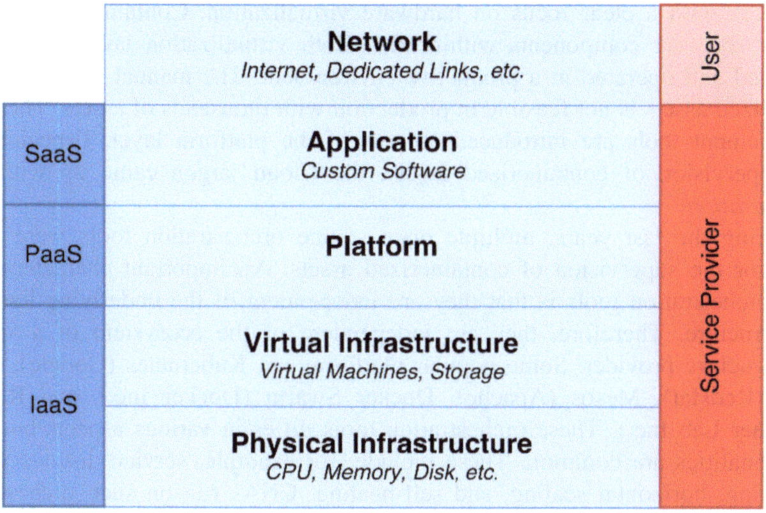

Fig. 3.1 Cloud computing model

platforms enables new stakeholders to use cloud infrastructure more efficiently. Cloud applications, which are meant to run on such platforms, require an adjusted design pattern that utilizes the provided Cloud infrastructure properly. Applications that apply with such design pattern are called Cloud Native Applications (CNAs).

Figure 3.1 visualizes the Cloud model [2]. The utilization of Cloud Computing requires an understanding of the different layers and their interaction with each other. The physical infrastructure layer contains the physical hardware and network resources. The NIST Cloud definition specifies four deployment models for Cloud Computing. They distinguish themselves by how many stakeholders share the physical hardware. They reach from publicly shared, over shared by a community of multiple stakeholders, to privately owned by a single legal entity. The last state is a hybrid of the other three. What deployment model is suitable for a use case depends mainly on security considerations and financial aspects. For a company, the operation of a private cloud setup requires the investment in a data center and a constant investment in personnel and maintenance. The operation of a private cloud system may be feasible for larger companies with a high demand on computational resources, or with high requirements on data security.

In a Cloud setup, the physical infrastructure is divided by the introduction of virtualization technologies. These technologies virtualize the physical hardware resources, like network, CPU, memory, and storage to make them easier manageable. In production environments, there can be hundreds or even thousands of such virtualized assets. There are various ways to set up a virtualization environment for private Cloud Computing. It is usual to have multiple virtualization technologies chained together. Commonly used are virtual machines (VM) and containers, which both serve different purposes. VM are usually used to manage and allocate hardware resources with a clear focus on hardware virtualization. Containers are used to deploy software components within VM. Both virtualization layers need to be managed and operated in a productive environment. The manual management of virtualized assets is not feasible in production with thousands of assets. Therefore, management tools are introduced as part of the platform layer. Especially for the supervision of containerized assets, the Cloud jargon came up with term *orchestration*.

During the last years, multiple open source orchestration tools were developed for the supervision of containerized assets. An important characteristic of such orchestration tools is that they are independent of the underlying hardware infrastructure. Therefore, they are independent of the ecosystem of a specific infrastructure provider. Some popular platforms are Kubernetes (Google), OpenShift (RedHat), Mesos (Apache), Docker Swarm (Docker Inc.), and Rancher (Rancher Lab Inc.). These orchestration tools differ in various aspects but some functionalities are common. These include, for example, service discovery, load balancing, horizontal scaling, and self-healing. CNAs run on such orchestration platforms. In order to exploit the full potential of the orchestration platforms a specific software design is required. Most platforms support a software design based on microservices. Microservices are isolated software components with dedicated memory that interact with their environment via Advanced Programming Interfaces (APIs) and messages. Most microservices are shipped containerized. A microservice is specialized to provide a specific functionality within the cloud application. The interaction of various microservices sum up to the full functionality of a Cloud application. More information about microservices can be found in [3].

Until now, we spoke about general software and application requirements for Cloud Computing and the purpose of orchestration frameworks. The available general purpose platforms provide the general framework for the development of CNAs. They are one side of the platform layer. Another side is the functional extension of the platform for specific needs of the application layer. Common examples are the support for domain specific data models or communication standards. In the energy sectors, a series of specific communication standards are used for different use cases. An example is IEC61850, a widely used standard for the monitoring of substation in the high to mid-voltage grid [1]. These application driven requirements can be used to develop more specialized application platforms. Such platforms are discussed in the following chapters to exemplify energy automatization.

Specialist platforms represent a basis for the development of Cloud applications. These applications are the backbone of new developments and use cases. Some of these use cases seem to have a fundamental impact on society. Such phenomenon is, for example, the ongoing digitalization of the physical world. This development is sometimes referred to as the Internet of Things (IoT) or in a more technical sense cyber-physical system.

3.3 Use Cases in Energy Automation: Monitoring, Control, and Management

Up to recent years, power distribution networks have been designed and operated as passive networks which are required to handle all probable loading conditions, resulting in over-dimensioned networks whose capacity is only occasionally fully used. There was no active network management, no real-time monitoring, and limited load demand shift day to night through retail tariff incentives. The only way to increase the number of serviced users was by expanding the network infrastructure.

Power distribution is undergoing fundamental changes towards more dynamic and complex structures, due to the integration of distributed generation (DG), distributed energy storage (ES), and new equipment and services, such as smart meters (SMs) and demand side management (DSM). In addition, the distribution network is becoming more meshed, with bidirectional power flows, and is changing from passive to active, requiring a new vision of active network management in order to support needed automation functionality such as monitoring and forecast of the status of the grid, control the distributed energy resources (DERs), and the grid equipment in order to reduce line congestions, address power quality issues, and improve overall efficiency.

The introduction of distribution grid automation functionality must be based on a system architecture framework and an analysis of the functionality that shall be provided. A framework that uses a cloud-based general-purpose system for distribution grids automation is presented below, effectively a supervisory control and data acquisition (SCADA) system. Analyses of the functionality that the SCADA system should support exist in literature as a subset of Smart Grid (SG) Use Case (UC) definitions performed available from EPRI. The EU project FINSENY[1] has also defined an extensive set of SG Use Cases.

In the specific area of DG automation, the IDE4L[2] project has explicitly addressed the design and deployment of a novel distribution network management automation architecture, based on a study of distribution grid Use Cases for monitoring, control, and business, in particular developing Intelligent Electronic Devices (IEDs) hosting substation automation functionality.

Offering a comprehensive vision of active distribution network integrated with active network management of resources such as DERs, MGs, and flexibility services, IDE4L is used here to exemplify certain automation paradigms.

IDE4L's automation concept encompasses:

- A hierarchical and distributed control architecture for distribution network automation;
- Virtualization and aggregation of DERs via aggregator;
- Large scale utilization of DERs in network management.

---

[1] *FINSENY* project (http://www.fi-ppp-finseny.eu/).

[2] *Ideal Grids for All* FP7 EEGI EU project (http://www.ide4l.eu) (FP7 EEGI).

The hierarchy includes three levels: primary (with autonomous controllers and protection devices), secondary (with cooperating IEDs in a network area), and tertiary (system level). IDE4L has not, however, considered the use of Cloud technologies or public communications infrastructures; these topics are addressed in Sect. 3.4 below.

3.3.1 What are the IDE4L Use Cases?

IDE4L has defined UCs in three clusters: business, control, and monitoring. The business use cases deals with creation and trading of energy and flexibility products. The control use cases are in the areas of power control, protection, power quality, and network planning. The monitoring use cases are:

- Real-time Monitoring (LV and MV): collection of measurements from IEDs in the substation and in SMs located at prosumers connection points;
- State Estimation (LV and MV);
- Power Forecast (LV and MV);
- State Forecast (LV and MV);
- Dynamic Monitoring: provision of dynamics information to TSOs based on PMUs;
- Network Description Update: update of databases following a network change;
- Protection Configuration Update: update of protection devices configuration following a network change.

3.3.2 What are the Communication Requirements of the IDE4L Use Cases?

The IDE4L project has defined a set of communication requirements for the UCs, using the classes defined in Table 3.1.

The transfer times specified for each UC must be fulfilled for the actual network configuration, including all the equipment and actors defined in the UC. In practice, there are dependencies between the transfer times and the transfer rates: for small network sizes, less data is produced and passed around between the actors, and it is consequently easier to meet the communication requirements than when the network size is increased. Any implementation of a UC must ensure that the communication requirements are fulfilled for the designed-for network size.

Some details of the communications requirements between the different IEDs in the IDE4L architecture are given for some of the UCs in Table 3.2. In LV Real-time Monitoring (LVRTM) real-time measurements are collected from the LV grid and stored in the Secondary Substation. In the control UC "Fault Location Isolation and Service Restoration (FLISR)," faults are located, the affected part of

Table 3.1 Communication requirements for distribution grid use cases

Transfer time class	Transfer time (ms)	Typically refers to ...
TT0	>1000	Files, events, log contents
TT1	1000	Events, alarms
TT2	500	Operator commands
TT3	100	Slow automatic interactions
TT4	20	Fast automatic interactions
TT5	10	Releases, Status changes
TT6	3	Trips, Blockings
Transfer rate class	Transfer rate (kb/s)	
TR0	<1	
TR1	1	
TR2	10	
TR3	100	
TR4	1000	
TR5	10,000	
TR6	100,000	

Table 3.2 Communication requirements for IDE4L use cases

Use case	Transfer time	Transfer rate	IEDs
LVRTM	TT6	TR6	Sensor → Substation IED
	TT0	TR6	Substation IED → Substation Automation Unit
	TT2	TR4	Substation Automation Unit → DSO SCADA
FLISR	TT3	TR6	Circuit Breaker → Substation Automation Unit
	TT6	TR6	Circuit Breaker → Actuator
LVPC	TT2	TR4	Field IED → Substation Automation Unit
DR	TT2	TR5	Service Provider Platform, Commercial Aggregator, DSO SCADA

the network is isolated by acting on circuit breakers and switches, and a service restoration process is carried out. Both LVRTM and FLISR have stringent real-time communications requirements. However, most other UCs do not need to operate in real-time or produce large data amounts: some examples shown are the control UC for "LV Power Control (LVPC)" that mitigates congestion in the LV networks, and the business UC "Day Ahead Demand Response (DR)" that determines a day-ahead least-cost demand plan.

3.4 Cloudification of Energy Automation Use Cases: Present and Future

3.4.1 Energy Services Platform

The way that power grids have traditionally been designed and operated is being revolutionized by the growing penetration of Distributed Energy Resources (DERs) into Low- and Medium-Voltage grids and by the increasing application of Information and Communication Technologies (ICT) in grid automation, resulting in the so-called Smart Grid. The balance between power generation by large, centralized power plants (such as coal-fired or nuclear) plants and power produced locally in distribution grids is changing in the direction of having more power being produced locally and fed directly into the distribution grid. This is a scenario for which power grids have not been designed, as the grids are laid out to support power being fed from the large power plants connected to the transmission grid to the loads in a top-down manner. Hence, widespread use of DERs is resulting in an ongoing revolution in power grids which effectively reverses the traditional top-down way that grids are laid out and operated, with centralized generation and centralized system-level automation. The Smart Grid is turning into a bottom-up system. This affects the business models of incumbent grid operators, who are used to being the sole providers of the commodity of electricity. In fact, it is a major game-changer that provides the opportunity for existing and new players to develop and implement new business models based on "energy as a service." The main driver behind these new services will be data, which has previously been the property of the incumbent operators and has been jealously guarded. It will be this data which will now carry the main value. Not just electricity but now also data will be traded and made available as a commercial product. Of course, making data available is a more open way must be done in a way that respects, and is compatible with, the right or individual persons to privacy. This means that the data will not be simply be traded in an uncontrolled way but will be subject both to legal restrictions and to commercial interests of the parties involved in buying and selling the data. In the end, however, the data will be made available in a form which agrees with the legal requirements and commercial interests to those parties entitled to use it and willing to pay for such use. These users can then use the data to flexibly define and implement customer services. In particular, this enables the incumbent Utilities to adopt business models which bring them closer to customers.

These services can be provided through a Platform which supports interoperability between the originators of the data and its users, while providing functionality for processing, storing, and adding value to the data. The energy domain has its own specific data and its own domain-specific needs, meaning that it makes sense to have a Platform providing services specifically for the energy domain, an Energy Service Platform (ESP). This is a software-based product or service that serves as a foundation on which outside parties can build complementary products or services, which are the Applications (in the following, referred to as "Apps").

3.4.1.1 What is an Energy Services Platform?

ESP is conceived as a cloud-based service platform for the Energy domain, applying the Service-Oriented Architecture (SOA) paradigm in order to provide a loosely integrated suite of services for commercial use to build value-added applications. ESP's purpose is to facilitate interoperability between these applications and devices in the field, i.e., in buildings or in the grid, providing cloud-based SCADA functionality for the Energy domain.

The Platform is an example of a cyber-physical system, enabling the different players to use common infrastructures for ICT, and data collection, processing, and storage, which reduces the cost of service provision. The Platform is not a monolithic entity but is heterogeneous as regards its

- *Sub-parts*: it can comprise many sub-platforms, or instantiations, each of which could support a different subset of Platform services, for example, in different countries with different regulatory systems;
- *Instantiations*: a given service could be offered in different Platform instantiations, for example, a service offering data on power plant production could be instantiated in different countries and offer data relating to the local power plants;
- *Technologies*: the services offered by the Platform are defined, but no particular Platform technology is mandated and new technological concepts can be incorporated; for example, cloud systems offered by different providers could be used for different instantiations of the Platform.

The technical approach shares several characteristics of other platforms reported in the literature and classified in Table 3.3:

- A service-oriented, open-source middleware, capable of supporting the business models of the different Smart Grid actors.
- A unified overview of data and a set of available services. This model decouples the logical representation of the Smart Grid domain from the physical infrastructure and enables interoperability between applications and devices through acquiring the data from various sources and giving authorized service providers easy access to it.
- Availability of Platform and component parts to framework partners only.
- Open-source concept, making the software components available to users, along with a software development kit (SDK), to speed up the development of new applications, letting businesses focus on what they best understand (energy), instead of ICT issues.
- Availability of basic services that support the prototyping of new applications.
- Option of decentralized instantiation (i.e., as an embedded platform in gateways or field devices) or instantiation in a centralized controller.

Being open source, the ESP:

- avoids being locked into a particular technology or vendor;
- encourages innovation,

Table 3.3 Classification of related work on platforms

Classification	Liakopoulos [4]	Volttron [5]	CENELEC [6]	Patti [7]	OS4ES [8]	Xin [9]	Pérez [7]	ENERGOS [10]
SOA approach	√	√	√	√	–	√	√	√
Unified view of data	√	√	√	√	NA	√	√	√
Basic services available	√	√	√	√	–	√	√	√
Software components available	–	–	–	–	–	√	–	–
Open-source approach		–	√		√	√		–
SDK available	√	–	–	–	√	√	–	–
Distributed/Centralized	DC	DC	C		D	D	DC	DC
Cloud based	–	–	–	–	–	–	–	√
Availability (Study, Prototype, Open-source, Commercial)	S	P	S	S	OS	OS	S	C

Key: "√" means "supported," "–" means "not supported." Aspects not addressed in study papers are indicated by leaving the cell empty

- allows development costs and results being shared by partners, so the benefit for the partners far exceeds that they would get from a technologically closed approach;
- allows adoption of new technologies and approaches as they emerge, thus ensuring that ESP is not locked into the technology that was available when it was designed, but can continue to benefit from global technological innovation;
- supports entry of new players by lowering entry barriers.

The ESP could run on a cloud infrastructure (in SaaS mode), or it could be implemented on a conventional dedicated computer infrastructure. To be able to take advantage of the possibilities offered by running on a cloud infrastructure, the ESP must be designed to allow the number of VMs on which it runs to be scaled up and down.

(In the rest of the text, we say "in the Cloud" to mean that we assume that the ESP runs on a cloud infrastructure).

3.4.1.2 What Does an Energy Services Platform Do?

The ESP provides application developers with a set of services (basic and, possibly, more complex services) implemented by a set of software components running in the Platform. New services can be created using the available existing components or by adding new components which offer some needed functionalities. New components can easily be added to the system. The components' interfaces are open and standard, which facilitates and promotes their re-use.

Examples of basic service are generic services, which are not related to the energy domain as such, like Data Aggregation or Complex Event Processing. It is envisaged that services of greater complexity and added value would typically be developed commercially as Apps. In the course of time, such complex services might be made available to the community as open-source and incorporated into the ESP. A pre-requisite for this is, of course, that the owner of the functionality agrees to the open publication of the service. It could be that the App has been further developed and its implementation split into a part that could be made openly available and a part which is still of commercial value and would be kept proprietary.

In order to ease the uptake of such Platforms, App developers should be offered an SDK with examples and pre-configured settings which make the services easy to use and thus foster quick app development preconfigured settings.

In case of "Build own ESP instance," ESP users can choose the open-source ESP components that they need and integrate them to build their own instance of the Platform to suit their individual needs. This instance can then be run on a cloud infrastructure of the user's choice. Pre-integrated and configured groups of basic components should be made available to make it easier to build ESP instances.

In the case where the user does not wish to build their own Platform instance, they could use an instance which may be offered on a commercial cloud platform in SaaS or PaaS mode.

3.4.1.3 Building on Ongoing Technology Developments

The principle of ESP is to re-use existing ICT concepts like Cloud Computing, Internet, and SOA so that the development of special solutions can be avoided and to incorporate future developments in ICT technologies.

Cloud Computing

Cloud Computing gives a distributed architecture that is not dependent on local hardware and supports data protection, consumer protection, and security through not having the data in one place [4]. It enables giving all actors access to all data while preserving confidentiality of competitors' data and privacy of end users. It supports parallel processing. SG applications require huge amounts of data to be stored: Cloud Computing offers scalability and cost-effective data storage, for example, through federation. Cloud Computing is globally accessible but can also be used to offer tailored services for a specific market, geographic or organizational area. There can be different platform instances with different sets of services, and it is capable of supporting industrial applications' needs on security, robustness, and availability.

Internet

The Internet is developing to be a place where a myriad of devices are interconnected, enabling new services to be defined and offered, building on improved communication technologies, IoT, Big Data, and cloud networking. Existing shortcomings of the Internet in terms of performance, reliability, scalability, security, and services are being addressed. This means that using the Internet will become increasingly appropriate for SG applications that need to handle huge amounts of data from many DERs. Hence, SG can ride on the coattails of future Internet development, benefitting from this technological evolution using it as a tool to enable the SG paradigm.

Many communication technologies that have previously had nothing to do with the Internet have been adapted to Internet technologies over the past two decades. Examples of this evolution are the use of Internet Protocol (IP) to carry both user traffic and signalling traffic in fixed and mobile telephony networks or the support of IP by Zigbee.

SOA

SOA allows simple, extensible APIs between the various actors, which hide underlying complexity, and allows distributed Cloud Computing to be used. An example of how SOA hides complexity is where services are offered whose implementation involves communication over a network.

SOA means, for example, that a grid operator can source needed grid data from the operator of the measurement equipment as a service rather than building duplicate measurement systems and special ICT. SOA has been applied in industrial automation applications.

Communications

Different communications technologies and standards are used in different SG UCs, depending on the purely communication-related requirements on latency and throughput but also on non-functional aspects such as security, reliability, robustness, availability, scalability, and Quality of Service (QoS). Communication is virtualized by the ESP, i.e., the users of ESP services receive communications as part of the service and do not have to worry about how the communication is realized. Hence, ESP is effectively offering a set of services which is overlaid on lower layers, such as communications, by means of which the services are realized.

Communications networks have experienced a process of maturation in the last decades which has seen the cost of communication being drastically reduced while the quantity of data communicated has mushroomed. Public telecommunications networks, both fixed and mobile, have been using Internet Protocols (IPs) and Internet technologies for more than a decade. Fourth generation mobile networks support Smartphone applications which require high data rates. It is to be expected that communication networks will continue to improve in terms of transfer times, transfer rates, and QoS. Future communication technologies may, however, support such requirements and future public communication networks may offer QoS classes supporting such applications. For example, 5G mobile networks, which will be rolled out in the next couple of years, will offer better support for IoT applications, featuring lower communication latency, and include support for huge numbers of devices, each sending and receiving small amounts of data occasionally. 5G will support edge computing, where communications network functionality and energy applications run in the radio base stations (called eNodeB) at the edge of the radio network, close to the distribution grid. These 5G mobile features will mean that 5G can be deployed in distribution grid applications requiring low latency, e.g., monitoring and control applications. Also, 5G mobile will offer end-to-end QoS classes, called QCI classes, suitable for energy-related traffic right down to the level of the individual devices in the field.

This means that use cases which are today not feasible to implement in the Cloud may become more feasible in the future. The technological tendency is towards the application of common communication and Cloud technologies across a range of industrial domains, which acts as a powerful driver for technological progress and will lead to Cloud systems being able to support more difficult UCs in the future.

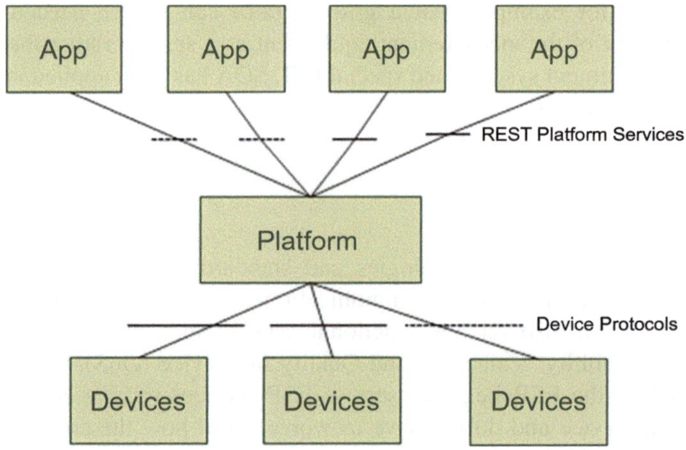

Fig. 3.2 Service platform for interoperability between field devices and apps

3.4.1.4 What Does the Energy Services Platform Architecture Look Like?

The general concept of a Service Platform is shown in Fig. 3.2. Its function is to provide interoperability between field devices and Apps. It supports various protocols towards field devices, gathering data from, and controlling, the devices. It provides a set of services which enable Apps to access the elaborated device data and to control the devices. The interoperability relates to the fact that a given App may access different device types and vice-versa.

ESP adopts a layered platform architecture, as shown in Fig. 3.3. The boxes shown inside the layers represent functional groups of software components or individual components. The SOA model of [5] is followed, in which horizontal layers contain the actual functionality and cross-cutting vertical layers provide the non-functional support which all the functional layers need. This non-functional support includes such things as security, support for inter-component messaging, information management services such as context awareness or entity registration, as well as QoS functionality. The components expose service interfaces to each other using the messaging service or externally (e.g. using Representational State Transfer (REST) or RESTful Web services or Microservices). Figure 3.3 does not show the underlying physical or virtual infrastructures which ESP runs on, e.g., databases, IT systems, communication services).

The Apps implement the added-value customer services, supporting the business of the application owner. The Apps are, in principle, external to the ESP. However, the Presentation Layer gives the possibility of running the Apps directly on the ESP.

The Service Logic Layer offers a set of services to the SG actors, supporting them in their roles and hiding the underlying complexity involved in implementing the services. It comprises middleware which integrates the data from the various devices and offers basic platform functionality through service APIs. It includes a database

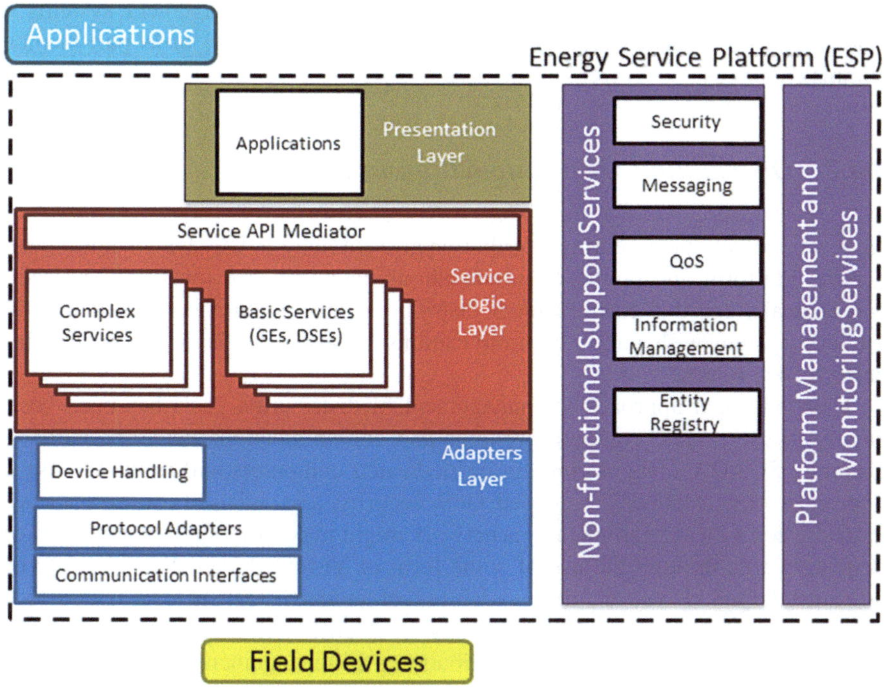

Fig. 3.3 Energy service platform architecture

which acts both as a data cache (if data is still available from the SE devices) or an archive (if the SE devices are no longer on-line). The middle layer incorporates routing to a module which executes the service (and determines whether to interact with the Adaptor layer and/or the database).

Device-oriented functionality (e.g., device registration) is supported by means of the Adapters Layer. It also terminates the communication interfaces between the Platform and the various field devices and information sources. It interprets the miscellaneous device protocols, mapping them to a common data model. It supports standardized protocols (e.g. Automation architecture IEC 61850, IEC 61599, OpenADR, etc.) but, if it is needed to access particular devices, could also support non-standard protocols. This means that the ESP must be able to deal with a number of different data syntaxes and semantics. Internally, the ESP maps these onto an internal meta-data model which unifies the underlying device data models based on a common semantic mode upon which the services APIs are based. The core asset of the ESP is, therefore, this *meta-data* model.

In order to achieve scalability of the Platform, gateways can be placed at the cloud edge to interface between local devices and the cloud, which perform functions like local control and data aggregation in order to reduce the amount of data being sent between the devices to the cloud platform. The division of functionality between cloud and gateway must be considered for each application.

The ESP can co-exist and co-operate with existing management systems or gateways, whose services can be offered to ESP users directly or as part of more complex services.

3.4.2 Distribution Grid Automation in the Cloud

Here, we look at implementing distribution grid automation UCs with a cloud-based Service Platform and public communications networks. For most of the UCs, it is technically feasible to implement distribution grid automation functionality in a Cloud system with public communications networks while fulfilling real-time performance issues.

The traditional approach to management of distribution grids has been based largely to use private computer systems and communications networks with, for example, SCADA systems running on dedicated computers and private LANs or fiber networks, to allow the needed QoS, security, and reliability to be achieved. Parts of such a DG communications network might make use of public communications networks through the use of some form of Virtual Private Network offering guaranteed QoS. However, the availability of modern ICT such as Cloud technologies and differentiated QoS offerings on Internet and mobile communication networks points towards a future distribution grid management scenario based on using public communication networks and Cloud Computing.

Cloud Computing can replace conventional remote terminal units (RTUs) and SCADA systems. Cloud Computing has already been applied to distribution grid automation to perform functions such as state estimation or grid monitoring. For example, the cloud-based state estimation system has a layer to interface to the field devices, a middleware layer to perform functions such as message broking, communications, device management and storage, and an Apps layer to implement added-value services. Another possible application of Cloud Computing is to replace computers in distribution grid substations.

The implementation of a Cloud-based approach to distribution grid automation is exemplified here using a simplified version of the DG Automation architecture of IDE4L, as shown on the left-hand side of Fig. 3.4, where Secondary Substation Automation Units (SSAUs) implement automation functionality for a DSO and provide data to the DSO's management systems (SCADA, DMS) and to other grid actors such as TSOs or Balance Responsible Parties. The right-hand side of Fig. 3.4 illustrates a Cloud system implementing distribution grid automation functionality. The functionality implemented in the Cloud includes the Substation Automation functionality and some SCADA functionality and can be fitted into a typical three-layer Cloud architecture. The Cloud implementation is shown here as being independent of any particular DSO: it interworks with data sources and Cloud-external management systems that could be owned by any grid actor. This illustrates that the application of Cloud technologies, if combined with data availability outside the data-originating organization (for example, via the REST interface shown in Fig. 3.4), can support the introduction of novel business models.

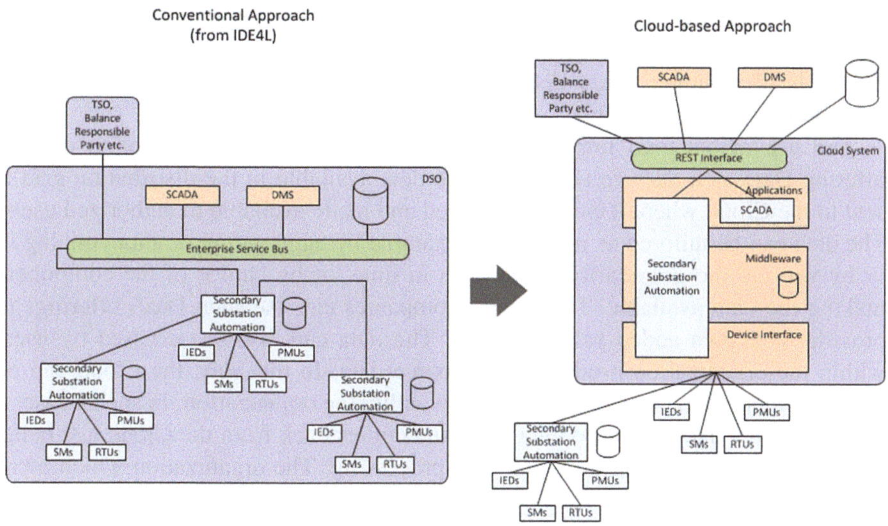

Fig. 3.4 Conventional distribution grid automation architecture mapped to cloud

The Cloud implementation retains the same interfaces between the SSAU, so that the Cloud System behaves like any other IED. The interface offered by the Cloud system to its users allows authorized and authenticated users to access cloud services; this is shown as being a REST interface, but other interface protocols could also be used.

The Cloud architecture shown in Fig. 3.4 can be flexibly applied to the particular application needs:

- It can be centralized or distributed;
- A private Cloud system can be made on an own physical infrastructure
- Otherwise, only certain functions could be implemented on the Cloud while retaining conventional implementations of other functions.

An advantage of Cloud implementations such as the above is that, if several different systems are implemented in the Cloud (e.g. the interface between the SSAU and the Cloud-based SCADA functionality on the right-hand side of Fig. 3.4), then interfaces that in a conventional implementation are between different physical systems become Cloud–internal interfaces. The same protocols can be applied in both cases but the communication requirements become much easier to fulfill, as the communication remains inside one system, the Cloud.

The limit of cloudification is that equipment that carries power cannot be removed from the physical distribution grid. Also, equipment that works directly with the power equipment (i.e. sensors and actuators) cannot physically be moved. Equipment used for grid protection is typically located close to the grid, typically in substations. It is physically feasible to move to the Cloud the part of such equipment that performs computation and control, subject to the communications between Cloud and grid being quick enough and the mentioned other factors being fulfilled. This seems quite an extreme scenario from today's standpoint, but cannot be discounted when one observes the continuing rapid developments in Communications technologies.

The use of a Cloud implementation of distribution grid automation enables offering Data as a Service (DaaS), where data available in the distribution grid is sent to the Cloud where it can be processed and made available to authorized users. The data can be tailored as needed, for example by aggregating or anonymizing it, or by varying the granularity of the data in time, or by choice of the component making the data available. Third party companies can use such DaaS offerings to provide their own added-value services. The data can also be accessed by users within the organization producing the source data. In this way, the Cloud can be used as a method to provide access to data within the organization. In such a case, it can be that the data is sent to the Cloud and comes back from the Cloud just being used for data storage, but doing no data processing. The organization which owns the source data and provides it as DaaS can charge for it and thus open up a revenue stream.

The pre-requisite for the use of Cloud Computing alongside public communications networks is that they offer the same level of non-functional performance in areas such as security, reliability, and resilience as implementations based on dedicated, special-purpose computing systems, and private communication. Because Cloud Computing and Communications technologies are being applied in a wide range of different industries, there is a great impetus in their development. This can be expected to lead to their increasingly widespread use by the distribution grid operators in the coming years, because they offer an experience equivalent to having one's own private network and computing systems at a fraction of the cost.

In conclusion, adopting a Cloud solution for distribution grid automation:

- Allows brings together data from different types of sources: substation automation, smart meters, weather, social events and makes the data available to many processes. This, in turn, enables the creation of smart data storage that can serve multiple goals of a company and integrate Operation, Planning, and Investment decision processes.
- Gives cost savings over non-Cloud IT solutions through being able to cost-effectively use (but not own) a scalable infrastructure and benefit from using latest technologies without having to implement them oneself, hence avoiding being locked into legacy technologies.
- Enables substation automation to be achieved with minimum infrastructure upgrade. It makes it unnecessary to add a full substation automation system. This is a major benefit because commercially available substation automation systems are not designed for to support distribution grids and their deployment there would be very costly.

3.4.2.1 Use of IEC 61850 in Cloud-Based Distribution Grid

The IEC 61850 communication standard [6] is widely used in distribution grid automation. It concerns communication for the purposes of sub-station automation between functions allocated to IEDs. The IEDs can be located at sub-station level, bay level, or process level. As a physical device (PD), it can contain one or more logical devices (LDs) hosting Logical Nodes (LNs), which are containers of the data communicated by functions and are defined in terms of basic data objects and attributes. In effect, IEC 61850 specifies an implementation-independent information model for communication between LNs and hence Cloud systems can directly be used as components in IEC 61850-based DG automation architectures. The pre-requisite for this is that the communications networks used must fulfill the real-time communications requirements.

To satisfy communications latency requirements, IEDs interfacing with Sensors and Actuators must be located close to the grid. However, the processing of data produced by such IEDs could be implemented in another PD. Conventionally, such PDs are located in substation automation equipment or SCADA systems, but may also be implemented in Cloud systems, as shown in Fig. 3.5. The internal structure of a Cloud system hosting an IEC 61850 compliant IED is different than conventional IEDs, but it has the same black-box behavior, defined by the LDs and LNs it hosts. Internally in the Cloud system, there could be several LDs implemented, as shown in Fig. 3.5, as the standard supports PDs containing several LDs. A Cloud system can, in one virtual system, implement functions that could be allocated to several different physical computers in a conventional implementation.

Hence, the IEC 61850 communications model can be directly applied to Cloud systems without the need to define new interfaces or protocols. Of course, any such Cloud-based implementation must also ensure that non-functional requirements such as reliability and availability are fulfilled.

Another aspect of putting IEC 61850 in the Cloud is that the communications technology used changes from being just Ethernet to also Internet or other technologies like mobile communications. In principle, any communications technology can be used as long as the communications requirements for the application are met. Also, the Cloud has not some of the characteristics normally associated with physical devices, such as a geographical location.

We can apply IEC 61850 in a cloud-based distribution grid automation architecture using public communications networks. This means putting distribution grid functionality into the Cloud. It results in a future substation architecture where the station bus is removed, the process bus connects IEDs with reduced local functionality and more functionality moved to the Cloud, and where IEDs concentrate data and communicates with the Cloud. The vision is of a virtualization of the distribution grid, meaning separation of grid management from the switching equipment, resulting in cost savings in equipment and operations. This future vision includes an end to the virtual integration of the distribution grid business with distribution grid operators, who do not own any grid equipment but offer grid management as a service.

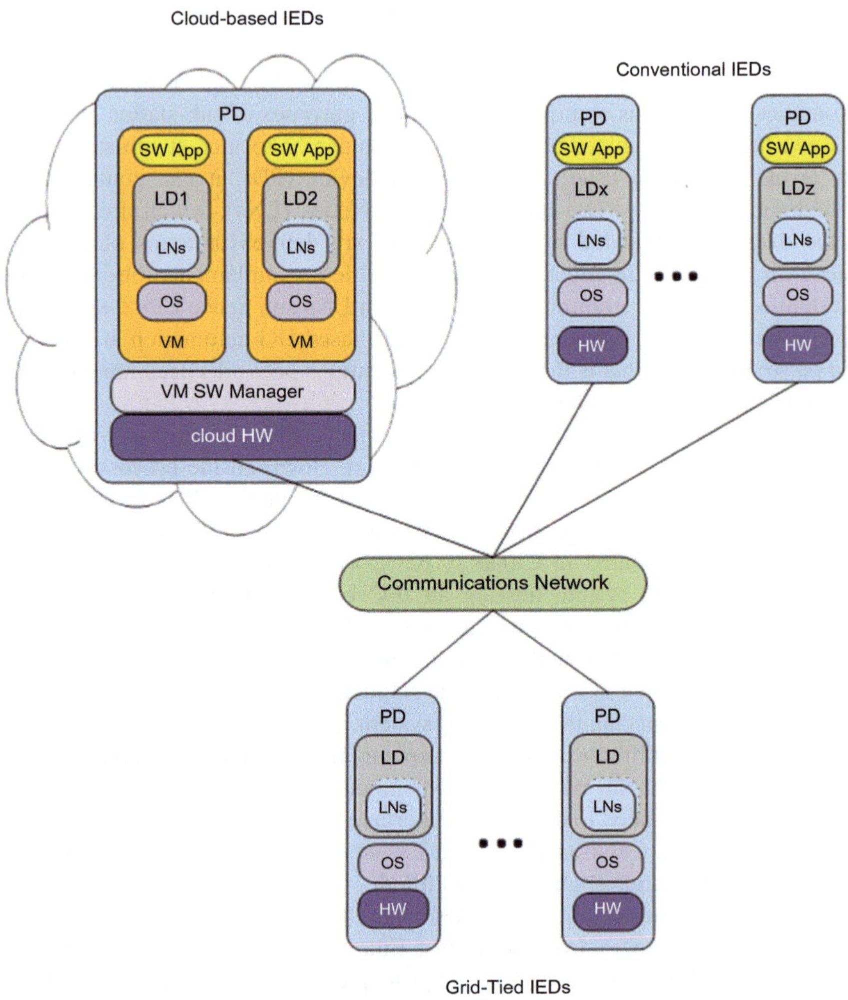

Fig. 3.5 Cloud-based IEDs versus conventional IEDs

3.4.2.2 Communication Requirements of Using Cloud Computing and Public Communications

Here we take the IDE4L UCs as an example, focusing their suitability for partial or full implementation in the Cloud, while fulfilling the transfer time requirements for communication between the actors over current public communications infrastructures. IDE4L has analyzed the requirements on information exchange in terms of transfer time and transfer rate and mapped them to a set of Communications technology classes, defined in Table 3.4.

Table 3.4 Technology classes (from IDE4L [1])

	Requirements			
Technology class	Transfer time	Transfer rate	Example technology	Cloud feasible?
Tech1	TT0, >1000 ms	TR4, 1000 kb/s	UMTS, BroadBand PLC on the MV grid, Broadband-PLC on the LV grid (cell with less than 50/100 of nodes), HiperLAN/Wi-Fi, FO	Yes
Tech2	TT0, >1000 ms	TR6, 100,000 kb/s	FO	Yes
Tech3	TT1, >1000 ms	TR3, 100 kb/s		Yes
Tech4	TT1, >1000 ms	TR7, 1,000,000 kb/s		Yes
Tech5	TT2, >500 ms	TR4, 1000 kb/s		Yes
Tech6	TT2, >500 ms	TR5, 10,000 kb/s	BroadBand PLC on the MV grid, Broadband-PLC on the LV grid (cell with less than 20/30 of nodes), FO, LTE, HiperLAN/Wi-Fi (with a point-to-point link)	Yes
Tech7	TT3, >100 ms	TR6, 100,000 kb/s		
Tech8	TT6, 3 ms	TR1, 1 kb/s		No
Tech9	TT6, 3 ms	TR2, 10 kb/s		No
Tech10	TT6, 3 ms	TR4, 1000 kb/s	FO, LTE (with no traffic), HiperLAN/Wi-Fi (with no traffic and with a point-to-point link)	No
Tech11	TT6, 3 ms	TR6, 100,000 kb/s	FO	No

Only some of the Communications technology classes are considered possible to support with today's technologies in a cloud implementation, i.e. where one of the IEDs which is communicating in the use case is located in a cloud-based ESP and the other is located outside the cloud, and where public communications networks are used. Of course, the achievable transfer time and transfer rate may depend on the actual communication network structure, the amount of other traffic being supported by the communication network, the size of the distribution grid that is monitored and controlled and the amount of data required to be transmitted.

TT2 class or higher (>500 ms) is possible to achieve using public internet. Disregarding the required transfer rate, this means that technology classes Tech1–Tech 6 are possible to support from a Cloud implementation, subject to the required transfer rate being supported, which is most difficult for Tech4.

An implementation between a Cloud infrastructure and a distribution grid infrastructure could use a Virtual private Network (VPN) or other mechanism to guarantee quality of service (transfer times and rates). A private Cloud locally in the Utility's premises using dedicated communication links could allow even TT6 to be achieved, as well as high transfer rates. It is considered that TT class requirements of TT3 or lower (100 ms latency or less) rule out a Cloud implementation unless special care is taken to design low latency communication methods.

The use of a gateway in the substation which communicates with the Cloud with a higher latency, together with a fast substation bus between the sensors/actuators and the gateway could provide an architecture using Cloud technologies in such a way that allows the requirements to be fulfilled.

The use of public Internet means that the achievement of a required transfer rate at a required latency is subject also to the general load on the network, which is variable. In general, reliability in meeting requirements on TT and TR can only be guaranteed by employing a network which supports QoS classes.

As regards the use cases themselves, then, while fulfilling the communication requirements given in Table 3.2 and fuller requirements given in [1], the majority of the UCs in IDE4L can be considered for a Cloud implementation. The exceptions are those with TT6 requirements of <3 ms latency. Such high communication data rates, <3 ms (TT6), occur in Real-time Monitoring (LV and MV) between the sensors and the IED, making these parts unsuitable for movement to the Cloud. Other UCs which also require very high data rates are the FLISR UCs, making movement of these to the Cloud difficult.

The business and control UCs do not have stringent communication requirements and could be implemented in a Cloud system. Indeed, such an implementation could result in interfaces between what are now separate systems becoming intra-Cloud interfaces, simplifying the communication.

The conclusion is that, because IEC 61850 defines Communications requirements in an implementation-independent way, any distribution grid automation function can be put into the Cloud, provided the communications requirements of the function are met, which is possible for most of the use cases with the public communications networks available today.

3.4.3 *Examples (Smart Metering, Demand Response, City Quarter Control, etc.)*

3.4.3.1 Moving the Secondary Substation Automation Unit to the Cloud

The SSAU is a computer component in the IDE4L architecture that implements monitoring and control functionality related to substation automation. It includes protocol handling (such as Manufacturing Messaging Specification (MMS) from IEC 61850, DLMS/COSEM), filters for bad data detection and extraction of meaningful information, and databases with different tables for several categories of data.

As shown in Fig. 3.4 above, the SSAU can be implemented conventionally, as was done in IDE4L, or as a cloud-based component. Whether implemented in the Cloud or conventionally, the SSAU is defined in IEC 61850 terms by the data produced by the set of LN(s) it implements, as defined by its Substation Configuration description Language (SCL).

The "Low Voltage Real-time Monitoring (LVRTM)" UC has been implemented in a demonstration Cloud system in order to prove the concept of Cloud-based implementation of the SSAU and the IDE4L distribution grid automation architecture in general [7].

LVRTM involves the acquisition of measurement data from Sensors in sub-stations in an LV network. The measurements are collected by sub-station IEDs (SSIEDs) and passed to the SSAU for further processing, involving

- Data Filtering, to discard any outliers of a given measurement point (due e.g. to tap changer operation or switching on/off of load or noise or missing samples).
- Data Harmonization, to perform harmonization in time between measurements with different reporting rate, providing an output adjusted to a uniform time window.
- Data Presentation, to perform visualization of data (raw and elaborated).

Above Table 3.2 gives an overview of the communication requirements between the Sensors, the substation IEDs, and the Cloud system. The required transfer times are >3 ms (from Sensor to IED), >500 ms (from substation IED to SSAU over MMS).

The required transfer time between substation IED and SSAU is achievable over today's public communication networks, allowing the SSAU part of the LVRTM functionality to be implemented in a Cloud-based system.

The demonstration system, as shown in Fig. 3.6 comprises two IEDs: an IED which produces distribution grid measurement data in IEC 61850 MMS format and a Cloud-based application which receives and processes the measurement data. The two IEDs communicate via public Internet using IEC 61850 MMS protocol. In the demonstration system, the measurement data were generated by a real-time grid simulation and passed over public Internet to the cloud system. The cloud system based on an Openstack implementation and internally used the Next

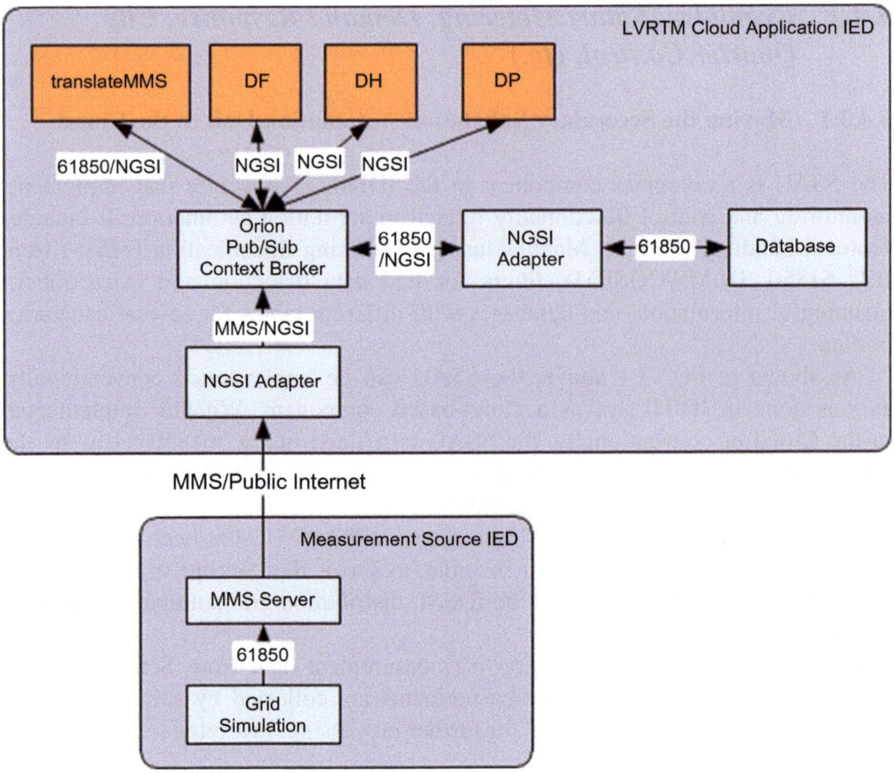

Fig. 3.6 Cloud Implementation of LVRTM use case

Generation Service Interfaces (NGSI) APIs [5] defined by the Open Mobile Alliance (OMA) for the development of services and Apps [5]. In particular, NGSI-9 is used to register context entities and NGSI-10 to subscribe to and retrieve context information. NGSI-9 and NGSI-10 are used for context handling by FIWARE's Orion Publish/Subscribe Context Broker GE.

In IEC 61850 terms, the measurement information is produced by the MMXU LN hosted in the substation IED and received by the MMXU LN hosted in the Cloud system.

3.5 Security and Privacy in Cloud-Based Smart Grids

A cloud-based SG can be conceived as a cyber-physical system that connects physical energy systems and cyber-infrastructure, with the integration of the internet. The SG can communicate with the consumers' appliances and also provide the service providers with the platform to acquire information and control operations. The presence of ICT and online connectivity inevitably exposes the SG to security and privacy challenges that can potentially disrupt the supply of energy services. Power theft by customers, data leakage, and data manipulation are among the most important security issues. To overcome these issues, we need to implement proper security actions.

The National Institute of Standards and Technology (NIST) Smart Grid Interoperability Panel classifies security in SGs into two categories: *cyber-security* and *physical security*. *Cyber-security* pertains to the information and networking infrastructure, whereas *physical security* refers to physical equipment, environment protection, and individuals (employees, staff, users) [8].

In this section, we present a synthesized overview of cyber-security and privacy aspects associated with Cloud Computing application to SGs, discussing objectives and requirements, the potential security threats (*cyber-attacks*) and the existing prevention and defense solutions (*attack countermeasures*). We also present an example of ongoing research & development in the field. Finally, we identify the uncovered problems that still need resolving, in both research and implementation.

3.5.1 Cyber-Security Aspects in the Smart Grid

3.5.1.1 Cyber-Security Objectives

The report released by the cyber-security working group in the NIST Smart Grid interoperability panel identifies three high-level SG security objectives [8]: availability, integrity, and confidentiality of data.

Data Availability refers to ensuring timely and reliable access to, and use of information, when it is needed. This means that the computing systems used to store and process the information, the security controls used to protect it, and the communication channels used to access it must function correctly. Ensuring availability involves not only preventing service disruptions due to power outages, hardware failures, and system upgrades, but also preventing *denial-of-service attacks (DoS)*, such as a flood of incoming messages to a target system resource in order to overwhelm it and make it unavailable to users.

Data Integrity refers to protecting against unauthorized, or undetected, modification or corruption of information, to maintain and ensure non-repudiation and authenticity of data over the entire life cycle. Information security systems typically provide message integrity in addition to data confidentiality.

Data Confidentiality means preserving authorized restrictions on information access and disclosure, primarily to protect personal privacy and proprietary information.

Availability and integrity are the most important security objectives in the SG, from the perspective of ensuring reliable energy service supply. Confidentiality is the least critical for system reliability, but it is becoming more important, particularly in systems involving interactions with customers, such as demand response and AMI networks.

3.5.1.2 Cyber-Security Requirements

Many organizations are working on the development of SG security requirements, including NEC CIP (North American Electrical Reliability Corporation—Critical Infrastructure protection), ISA (International Society of Automation), NIPP (National Infrastructure Protection Plan), IEEE 1402, and NIST. Based on the NIST report and existing research on cyber-security, we identify three areas with main security requirements.

Attack Detection and Resilience Operation

Since the SG features a relatively open communication network over large geographical areas, it is very critical to maintain any SG part or node invulnerable to network attacks. Adopting robust countermeasures for attack detection and identification is essential, and these countermeasures must be widely deployed everywhere. Moreover, the SG must have the self-healing ability to continue the network operations in the presence of attacks. Due to the critical importance of power infrastructures, resilience operations in the communication network are essential to sustain the availability of the SG infrastructure.

Identification, Authentication, and Access Control

The SG infrastructure can integrate up to millions of electronic devices and users. *Identification and authentication* is the key process of verifying the identity of a device or user, as a prerequisite for granting access to resources in the SG information system. *Access control* ensures that resources are accessed only by authorized and correctly identified users. Strict access control must be enforced to prevent unauthorized users from accessing sensitive information and controlling critical infrastructures. To meet these requirements, every node in the SG must have at least basic cryptographic functions, such as symmetric and asymmetric data encryption.

Secure and Efficient Communication Protocols

Unlike conventional networks, in SGs the message delivery must be timely and secure. Time-criticality and security, however, can be conflicting requirements: since networks (or sub-networks) in the SG cannot always use secure, physically protected, and high-bandwidth communication channels, optimal tradeoffs are required to balance communication efficiency and information security in the design of communications protocols and architectures for the SG.

3.5.1.3 Network Security Threats

Cyber-security threats place the SG components at risk, especially when the status of the point of common coupling (PCC) or the control orders for managing the different agents are altered. Since security issues mainly come from malicious cyber-attacks via communication networks, it is essential to understand potential vulnerabilities along the entire extension of the ICT infrastructure of the SG, under network attacks.

Attack Classification

In communication networks, security attacks can be classified into two categories: selfish misbehaving users, and malicious users, both posing severe security problems to communication networks. *Selfish misbehaving users* attempt to obtain more network resources than legitimate users by violating communication protocols. In contrast, *malicious users* aim to illegally acquire, modify, or disrupt information in the network.

In SGs, malicious behavior can have more serious implications than selfish misbehavior. Since a large number of electronic computing devices are used for monitoring and control purposes, and not only for simple functions like in the Internet (e.g., data services supply, file downloading, data sharing, etc.), malicious attacks could cause catastrophic disservice in power supplies and widespread power outage, which is an unacceptable case.

Therefore, we choose to focus on malicious attacks, classifying them into three types based on the SG security objectives, that is, availability, integrity, and confidentiality. Attacks targeting availability can be either *denial-of-service (DoS)* attacks, or electrical power attacks, or *server room environment attacks*. *Attacks targeting integrity* attempt to modify or disrupt data exchange. *Attacks targeting confidentiality* aim at acquiring unauthorized information from network resources. Unlike attacks targeting availability, which can occur at various layers, attacks targeting integrity and confidentiality are typically launched at the application layer, since they aim at acquiring or manipulating data.

Denial-of-Service Attacks

DoS attacks [9–11] attempt to delay or block communication in the SG, to compromise the operation of electronic devices. They can happen at the application layer or at higher communication layers in the SG. They can range from stronger attacks, where the attacker aims at completely disrupting the access to network or resources, to weaker attacks, where the attacker aims at violating the timing requirements of a critical message exchange. The latter case can also be catastrophic for the power infrastructure. For example, a DoS attack can severely degrade power equipment if it successfully delays the transmission of a protection message in the case of trip protection in substations. A common DoS attack is the Internet Control Message Protocol (ICMP) attack, also known as a Smurf attack. The attacker sends error messages to indicate issues that occur between networked entities. Since ICMP is a required part of the Internet Protocol (IP), and the modern power grid is IP-based, most devices, such as network enabled DC power and energy meters, accept the ICMP request, which can cause disruptions in service and maybe even a system crash.

Distributed Denial of Service (DDoS) attacks are amplified DoS attacks, originating from many attacking machines (*zombies*) from different geographical regions (*zombie network*).

Electrical Power Attacks

These attacks target the availability of the individual power components and can lead to power loss, unwanted voltage/current fluctuations, etc. Since new devices are constantly integrated in the system and deployed, the potential sources and targets of vulnerabilities in the system undergo constant evolution. Standards and regulations for those new components and their compatibilities need to be modified accordingly.

Server Room Environment Attacks

These attacks target the environment of the server room of the ICT infrastructure (fire, temperature, water, humidity, etc.).

Attacks Targeting Data Integrity

Attacks targeting data integrity [12] are unauthorized attempts to modify data in order to corrupt the exchange of critical information in the SG. The target can be either customers' information (e.g., pricing information, account balance, etc.) or state estimates of power systems (e.g., voltage readings, device running status, etc.). Because such information in power systems is valuable to both end-users and utilities, fault-tolerant methods and integrity checks are widely deployed in SGs to protect data integrity. However, since the number and type of integrity threats steadily rises, these countermeasures need continuous revision and upgrading. For instance, *false data injection attacks* are a potential class of cyber-attack that was initially designated against state estimation in SCADA system and then proven to be possibly extended against state estimation in electricity markets, or against load bus injection measurements (*load redistribution attack*s).

False data injection in electricity market [13] would circumvent the integrity checks that are used in present SCADA systems and lead to profitable financial misconduct. *Load redistribution attacks* [14], instead, could successfully bypass bad data detection and manipulate the state estimation outcome, leading the security-constrained economic dispatch of the SG to a false secure and optimal operating state.

Attacks Targeting Data Confidentiality

Compared to attacks targeting integrity, *attacks targeting confidentiality* do not modify information transmitted over power networks. They hack communication channels in power networks to acquire confidential information, such as a customer's account number and electricity usage. Typical examples include [15]

- *password attacks*, used to hack the passwords of users of a target computer to gain access;
- *packet sniffing*, where the attacker capture data packets (typically Ethernet frames) in travel, in order to read sensitive data like passwords or card numbers, if the network traffic is not encrypted;
- *port scanning*, where the attacker attempts to discover the services running on a target computer by scanning the TCP/UDP ports.
- *dumpster diving*, that is to say, searching through company dumpsters for any information that can be useful for an attacker for attacking the network (employee names, software application product information, network infrastructure, device make and models, etc.)
- *wiretapping*, where the attacker hacks the telecommunication devices listen to the phone calls of others.
- *phishing,* an attempt to hack sensitive information (usually financial information like bank user ID/password, credit card details, etc.), by sending unsolicited emails with fake URLs.
- *pharming*, aimed at redirecting the traffic of one website to another website.
- *social engineering,* a type attack in which someone with very good interactive skills manipulates others into revealing information about network that can be used to steal data

Although data confidentiality attacks have negligible effects on the functionality of communication networks in the SG, nevertheless their social impact can be severe. Due to the increasing awareness and importance of customer privacy, the social impacts due to confidentiality attacks receive more and more attention in present years, especially considering the potential leakage of massive customer information.

3.5.2 *Privacy Policy and Requirements*

Privacy problems are inherent in the SG. The sophisticated SG infrastructure is designed to enable interoperability between energy service providers and consumers, for the sake of intelligent control and economic management of energy consumption. Because the exchange of information lies at the heart of the interoperability paradigm, the implementation of the SG can have significant implications for personal privacy. Therefore, energy service providers and regulators must take great care not to sacrifice consumer privacy in the process of developing and implementing the SG. For example, providers must ensure that all critical personal data (energy consumption readings, energy bills, credit card numbers, etc.) are masked or encrypted and that only authorized users have access to data in its entirety. Moreover, digital identities and credentials must be protected as should any data that the provider collects or produces about customer activity in the cloud.

In the following, we give an overview of the possible privacy issues.

3.5.2.1 Personal Information

Personal information is any recorded information that can identify an individual directly or indirectly. In the SG context, any energy use data could link to personal information. NIST report provides the following list of personal information that could be available in the SG:

- *Name*: account holder
- *Address*: location where service is provided
- *Account number*: unique identifier of the account
- *Meter readings*: record of kW and kWh consumption at 15- to 60-min interval during the current billing cycle
- *Current bill*: current amount due to on the account
- *Billing history*: past readings and bills, including history of late/failed payments
- *Home Area Network (HAN) and in-home electrical appliances*
- *Lifestyle*: home occupancy, consumption patterns, etc.
- *DER*: *the* presence of on-site generation or energy storage, their operational status, net supply to or consumption from the grid, usage patterns
- *Meter IP*: IP address of the meter
- *Service provider*: identity of the service supplier (relevant only in retail access markets)

3.5.2.2 Privacy Concerns

NIST identifies four typical areas of privacy concern in the SG, as follows [16]: (1) fraud; (2) data in the smart meter and HAN; (3) near-real-time data; (4) personal lifestyle.

Energy fraud is any intentional tampering actions on the metering system carried out by the consumers or their agent, so that the record of the consumption is lower than the actual. Energy scams are also a form of potential energy fraud. To prevent this, robust fraud detection methods should be deployed, while preserving the privacy of consumers.

Data in the smart meter and HAN could reveal the appliances used and/or types of activities, and track specific times and locations of energy consumption in specific areas of the home. Appliance vendors may want this kind of data to know what customers use their products, when and how, and use such information to revise appliance warranties. Other entities may use this data to conduct target marketing.

Obtaining *near-real-time data* about energy consumption could indicate whether premises are occupied, where people are in the premises, what they are doing, etc.

Personal lifestyle information derived from energy use data could be valuable to some sellers or parties. Vendors could use this information for targeted marketing, and this could be not welcomed by these targets. This information could be revealed by SMs, time of use and demand rates, and direct load control of equipment. It could be forwarded or sold to third parties, and used for energy management analysis and peer comparisons.

3.5.2.3 Privacy Recommendations

The NIST report [16] provides a consumer-to-utility privacy impact assessment of the smart grid, proposing ten potential design principles to address privacy issues, as follows.

1. Organizations should ensure that information security as well as privacy policies and practices exist, are documented and are followed.
2. Before collecting and sharing personal information and energy use data, service providers should clearly notify their customers.
3. Service providers should give users multiple choices of service.
4. Organizations need to obtain users' consent or implied consent if it is not feasible, with respect to the collection, use, and disclosure of their personal information.
5. Organizations should collect from individuals only personal information that is necessary to fulfill the stated purpose. Treatment of the information should conform to the privacy principles.
6. Information should only be used or disclosed for the purpose for which it was collected and should only be shared with authorized parties. Personal information should be aggregated or anonymized wherever possible. Personal information should only be kept as long as is necessary to fulfill the purposes for which it was collected.
7. Organizations should allow individuals to check their personal information and to request correction of perceived inaccuracies.
8. Personal information should be protected from unauthorized modification, copying, disclosure, access, use, loss, or theft.
9. Organizations should ensure completeness and clarity of information regarding data usage.
10. Privacy policies should always be made available to service recipients.

3.5.3 SUCCESS Project

In the specific area of SG cyber-security, the ongoing R&D project "SUCCESS"[3] proposes an overarching approach to analyze security threats and countermeasures, with specific focus on vulnerabilities introduced by SMs.

Using a security-by-design approach focusing on resiliency and survivability, SUCCESS aims at designing an Energy and ICT Infrastructure, based on a New-generation Open Real-time Smart Meter (NORM) capable to secure the end nodes of the power system while providing innovative services in a customer centric grid.

At systems level, the project proposes a cloud approach, based on double virtualization, aimed at providing:

- Security countermeasures
- Security Monitoring Centers at DSO and Pan-European levels
- Secure communications solutions using Network Function Virtualization (NFV) and the LTE and 5G mobile communication technologies.

This work is going to be complemented by data privacy studies to ensure the acceptability of the results by consumers.

Trials are run in Ireland, Italy, and Romania.

3.5.3.1 SUCCESS Scope

SUCCESS aims at the design, development, and validation on small scale field trials of a novel holistic adaptable security framework to reduce the risks for additional potential cyber threats and attacks when next generation real-time scalable unbundled smart meters are deployed along smart electricity grid.

The project addresses both utilities networks and the communications networks and IT capabilities that support them. It considers both these infrastructures as they are currently used, as they will be used in 2020 with Next Generation functionality, and finally as they will be used and designed in the time beyond 2030 (Fig. 3.7).

SUCCESS examines the interdependencies and exploits the potential interactions between the communications and critical utility infrastructures in order to use them for countermeasures, embedding security, resilience, and survivability to Future Smart Grid Networks. *Security* includes security of supply and resilience to all types of threats: not only cyber-attacks but also weather or any other potential disruptive events such as loss of equipment.

[3]SUCCES project (http://www.success-energy.eu) (EU Horizon 2020 Research and Innovation Programme).

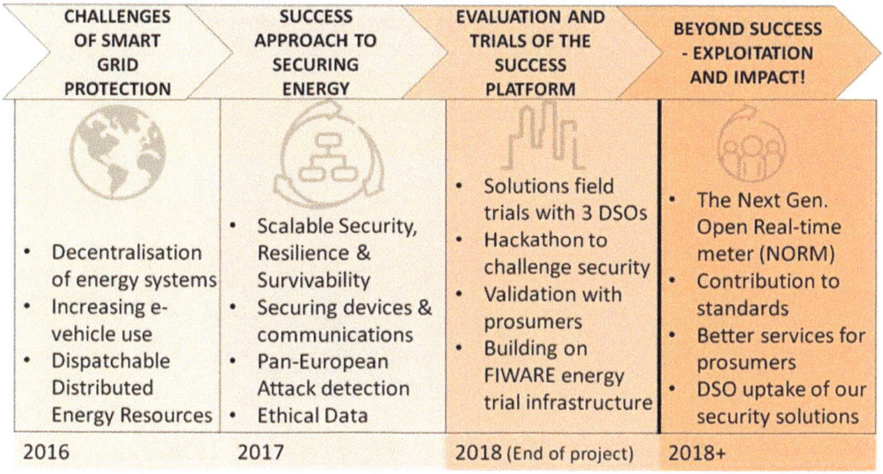

Fig. 3.7 SUCCESS timescales

3.5.3.2 SUCCESS Architecture

The SUCCESS architecture, as shown in Fig. 3.8, encompasses:

- Components providing security functionality from meter or smart device at a range of levels right up to the Pan-European security monitoring center level;
- Communications based on mobile 4G and 5G technologies, including low latency;
- Service APIs provided by a cloud-based middleware platform which interworks with devices and gateways in utility's grids;
- Security applications, such as the European-level Security Monitoring and Information System (E-SMIS), which use these service APIs;
- New-generation open real-time smart meter (NORM)

Over the years, SUCCESS will produce a comprehensive framework for securing SGs and similar critical infrastructures expressed as:

- A set of concepts for Security, Resilience, Survivability, and Privacy by design;
- Further development of the NORM produced by the NobelGrid project,[4] with security functionality and the inclusion of a Phasor Measurement Unit (PMU);
- A range of prototypes and blueprints for future energy and ICT sector products and services;
- Recommendations on countermeasures to address short-, medium-, and long-term threats;

[4]NobelGrid project (www.nobelgrid.eu).

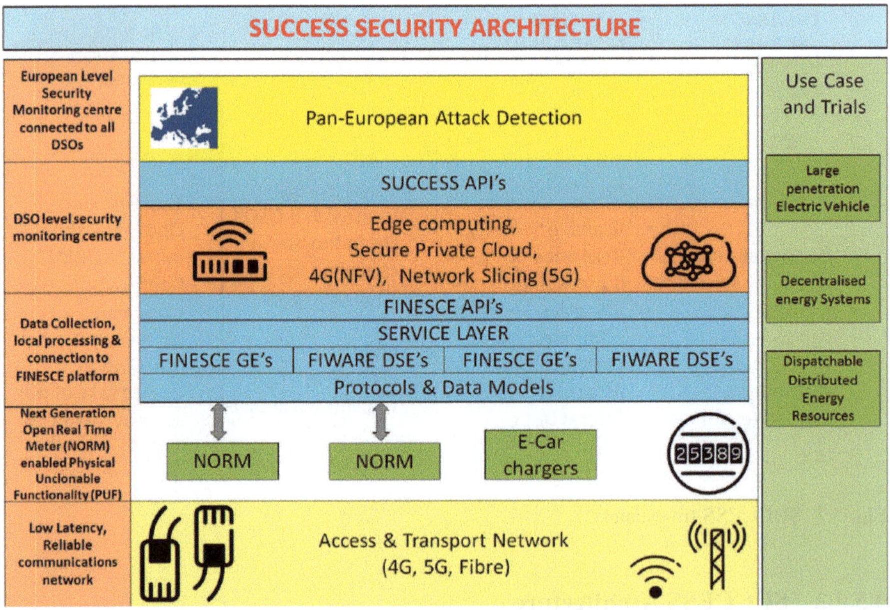

Fig. 3.8 SUCCESS architecture

- The SUCCESS platform implementing countermeasures and demonstrating in field trials Specifications for Certification testing and standardization of the approaches.

3.6 Conclusions

The development of Cloud Computing in the distribution grids automation sector is strictly dependent on the parallel evolution of the extremely large, dynamic, and substantially independent ICT and Energy markets. Both markets are global and highly articulated, and the complexity of challenges that accompany their development is further amplified by the prospect of using the Internet. This could, on one hand, create great opportunities for employing new products and services, but on the other hand, aggravate emerging critical issues in operation reliability, information protection, and data privacy. The difficulty of deploying Cloud Computing in automation is also linked to the emerging of a global, Internet-based, communication system that offers, with its incredible potential, expansion of performance, benefits, and services. Problems arise especially in terms of "shadow" costs linked to risks and to possible physical or virtual damages related to a non-complete control of data and of related processing of information. The major risks are in the processes that are most vulnerable or more sensitive to techno-economic value and to security (e.g., protection coordination mechanisms, data acquisition, processing and measurement, support services for control centers and operational management, information sources supporting single or multiple aggregation of market operators).

As technological progress and markets cannot stop, on the other hand, the benefits of Cloud Computing techniques within distribution system automation will be useful in practice only if the philosophy of application meets the requirements of market and users (i.e., energy service providers and customers). So, both ICT and Energy sectors will benefit from research, industrial experimentation, and development, aimed at defining applications and models of use in terms of system configuration and control architectures, and at providing proper application interfaces to the various network components. Complementarily to ICT application interfacing, it will be necessary to clarify the political and commercial aspects of the Internet infrastructure as a global communication network, and to systematize and regulate its use at the different levels of market deregulation.

References

1. Mell P, Grance T, The NIST definition of cloud computing, Recommendations of the National Institute of Standards and Technology; 2011.
2. Moghaddam FF, Rohani MB, Ahmadi M, Khodadadi T, Madadipouya K, Cloud computing: vision, architecture and characteristics. In: IEEE 6th control and system graduate research colloquium; 2015.
3. Newman S. Building microservices. Newton, MA: O'Reilly Media; 2015.
4. The Open Group. SOA reference architecture. 2011 [Online]. Available: http://www.opengroup.org.
5. Next Generation Service Interfaces Requirements. Approved version 1.0, open mobile alliance OMA-RD-NGSI-V1_0–20120529-A; 2012.
6. International Electrotechnical Commission (IEC) Standards. IEC61850: Communication networks and systems for power utility automation. Available: https://webstore.iec.ch/ publication/6028
7. McKeever P, Monti A. NIST Special Publication 1108R2. NIST framework and roadmap for smart grid interoperability standards. Release 2.0; 2012.
8. NIST Special Publication 1108R2. NIST framework and roadmap for smart grid interoperability standards. Release 2.0; 2012.
9. Zhang P, Wang H, Hu C, Lin C. On denial of service attacks in software defined networks. IEEE Netw. 2016;30(6):28–33.
10. Yan Q, Yu FR, Gong Q, Li J. Software-defined networking (SDN) and distributed denial of service (DDoS) attacks in cloud computing environments: a survey, some research issues, and challenges. IEEE Commun Surv Tutorials. 2016;18(1):602–22.
11. Yan Q, Yu FR. Distributed denial of service attacks in software-defined networking with cloud computing. IEEE Commun Mag. 2015;53(4):52–9.
12. IEEE draft trial use standard for a cryptographic protocol for cyber security of substation serial links. In: IEEE P1711/D7; 2010. p.1–41.
13. Xie L, Mo Y, Sinopoli B. False data injection attacks in electricity markets. In: 2010 First IEEE international conference on smart grid communications, Gaithersburg, MD; 2010. p. 226–31.
14. Yuan Y, Li Z, Ren K. Modeling load redistribution attacks in power systems. IEEE Trans Smart Grid. 2011;2(2):382–90.

15. Fan Z, et al. Smart grid communications: overview of research challenges, solutions, and standardization activities. IEEE Commun Surv Tutorials. 2013;15(1):21–38.
16. The Smart Grid Interoperability Panel – Cyber Security Working Group. Guidelines for smart grid cyber security, NISTIR 7628; 2010.

Chapter 4
Privacy-Preserving Smart Grid Tariff Decisions with Blockchain-Based Smart Contracts

Fabian Knirsch, Andreas Unterweger, Günther Eibl, and Dominik Engel

4.1 Introduction

A global trend towards the modernization of energy grids is ongoing: Information and communication technologies are integrated into the energy grid infrastructures, creating so-called "smart grids." A multitude of new use cases become possible, including the balancing of power generation and consumption, electric mobility, renewable energy sources, and real-time pricing. The modernization on the technical side is accompanied by a move towards deregulation on the regulatory side [1], thereby vastly increasing the number of choices on the side of the end consumer. In liberalized energy markets, the paradigm for the customers will be changed: They will be able to change tariff, provider, or both frequently and to adapt dynamically. There are a number of pilot regions that have shown the benefit of this deregulation, e.g., the German *eEnergy* projects, most notably the *MeRegio* project [1].

Customers have to choose one tariff out of those offered from their local utilities. In order to select the optimal tariff, i.e., the one fitting their electricity consumption at a given time, customers need to match their current usage habits to the offered pricing schemes. As an example, consider an energy provider which is interested in moving energy consumption away from times with high demand because this decreases the amount of immediately available but expensive energy resources. In order to encourage customers to use energy at off-peak times, the provider could offer better tariffs for load profiles that are dissimilar to normal consumption profiles by, e.g., lowering a fixed-price tariff from 23 cents/kWh to 17 cents/kWh. As another example, in order to stimulate people charging their cars during the night instead of during the day, a load profile with high consumption during night might be

F. Knirsch • A. Unterweger (✉) • G. Eibl • D. Engel
Josef Ressel Center for User-Centric Smart Grid Privacy, Security and Control, Salzburg University of Applied Sciences, Urstein Süd 1, 5412 Puch bei Hallein, Austria
e-mail: fabian.knirsch@en-trust.at; andreas.unterweger@en-trust.at

W. Rivera (ed.), *Sustainable Cloud and Energy Services*,
DOI 10.1007/978-3-319-62238-5_4

associated with another cheap tariff. A convenient way to perform the matching of current and desired consumption profiles is for the customer to provide a current load profile, i.e., a recent energy usage record, to different energy providers, to a third party, or a blockchain-based smart contract providing a matchmaking service for tariffs offered by various energy providers.

However, while this is a convenient approach, there are privacy concerns: It has been shown that personal information, such as lifestyle, religion, habitual patterns, sleep-wake-cycles and activities can be extracted from load profiles (e.g., [2, 3]). On the side of the energy providers, there is also reluctance to disclose the load profiles that different tariffs are based on (e.g., typical load profiles of customer groups), as this is internal information that, in a liberated market, can make a difference in commercial success [1]. Therefore, while both, the consumer side and the provider side, are interested in facilitating optimal matching, the aforementioned are strong reasons against disclosing full information for the matchmaking process.

In [4], a method for tariff selection has been presented that preserves both, consumer privacy and the internal company data of energy providers. This is achieved by the use of embeddings where load profiles are transformed into a domain that does not allow to reconstruct the original load profiles, but preserves the distances. The method outperforms previously suggested methods. In particular, it limits data expansion, which is an issue when classical homomorphic encryption is used. The trade-off is matching accuracy, which in this method is not 100%. Depending on choice of parameters, higher accuracy can be traded for increased data expansion. However, in real-word application scenarios, the accuracy of 93.5%, which can be realized with the proposed approach at negligible data expansion, is sufficient.

In this paper, we propose an adaption of the approach in [4]. Our proposed extension of this approach uses blockchains and smart contracts [5, 6]. While the original approach requires a third party for determining the minimum distance and the optimum match, respectively, we propose omitting the third party and replacing it with a fully decentralized and trust-less[1] network. Smart contracts allow every peer—customers and utility providers—to calculate and verify the results. This improves the existing solution in terms of verifiability, reliability, and transparency.

The possible use cases of the proposed protocol are manifold. An exemplary use case would be for a number of utilities to (continually) analyze and cluster the energy usage data of their customers. Based on this analysis, different tariff groups can be created. Consumers who wish to select the tariff which is best suited to their current consumption habits can create a smart contract with their load profiles and commit it to a blockchain where both, the consumers and the utilities are peers. The utilities then commit a number of load profiles that represent tariff options and the smart contract evaluates the best-matching tariff. The result will be verifiable by all parties.

[1]The trust-lessness of blockchains relies on the assumption that at least half of the computing power is spent by honest peers [5], as will be described in Sect. 4.3.3.1.

Note that during the whole process, no load information is disclosed by either the consumer or the utilities, i.e., the other peers cannot access the consumer data as it may reveal sensitive information about them [7, 8]. The tariff selection is then done decentralized and trust-less by executing the smart contract in the blockchain.

In summary, this paper extends the previously presented approach in [4] based on nearest neighbor embeddings [9] and oblivious transfer [10] for finding the best-matching tariff without directly revealing any load profiles to any involved party. This paper contributes (1) a fully decentralized and trust-less approach for tariff-matching; and (2) an exemplary implementation as a smart contract. The proposed approach is designed for the preservation of privacy, i.e., as in [4], privacy relies on the security of embeddings [9]. While the privacy guarantees of the proposed approach are not as strong as the ones in [4], a way to mitigate this is outlined.

The rest of this paper is structured as follows: In Sect. 4.2, related work from the domains of secure distance computation, blockchains and smart contracts, and other related approaches is discussed. Section 4.3 introduces the preliminaries such as design goals, the original profile matching protocol with a third party and blockchain-based smart contracts. In Sect. 4.4 the decentralized and trust-less protocol for tariff-matching based on a blockchain is presented in detail. Section 4.5 evaluates the proposed protocol, presents a conceptual implementation of the smart contract, and compares the proposed protocol to related work. Section 4.6 summarizes this work and gives an outlook to future research.

4.2 Related Work

In this section, we provide an overview of related work. We distinguish between secure distance computation methods which are suitable to compare load profiles, existing work related to blockchains and smart contracts and other related work including oblivious transfer and smart-grid literature with tariff selection and profile matching.

4.2.1 Secure Distance Computation

The following related work describes secure distance computation algorithms. In our approach, we compare load profiles using Euclidean distance measures, which is a similar type of computation.

Mukherjee et al. [11] propose a method where a number of selected Fourier transform coefficients are permuted and communicated between the involved parties. The properties of the Fourier transform are used to provide limited guarantees on distance preservation based on the selection of transform coefficients. This makes their approach probabilistic with the number of coefficients retained providing a trade-off between privacy and accuracy. Although our approach is probabilistic as well, the level of privacy is not influenced by the parameters which affect the accuracy of our approach.

Ravikumar et al. [12] describe an algorithm for secure Euclidean distance computation. Their approach is probabilistic as well, but requires a high number of vectors to be compared in order to come close to the actual distance values. In contrast, our approach finds the minimum distance in nearly all cases, allowing for near-perfect matching.

Wong et al. [13] introduce transformations of database points and query points that enable a ranking of the database points with respect to their nearness to the query points suitable for k-nearest-neighbor determination. However, their transformations are not distance-preserving which enhances security against attackers with known plaintexts, but limits possible applications.

Rane et al. [9] and Boufounos et al. [14] propose the use of distance-preserving embeddings for privacy-preserving matching and nearest-neighbor searches in the context of image retrieval. We apply these distance-preserving embeddings as one step in our proposed protocol and make use of their privacy-preserving properties.

Homomorphic cryptosystems can be used for secure distance computation [15, 16], e.g., in the context of image retrieval [17], fingerprint matching [18], and face recognition [19]. Kolesnikov et al. [20] combine homomorphic encryption for computing distances with Garbled circuits for choosing the point having the minimum distance to the query point. We provide a detailed comparison between our original approach, the approach based on blockchains and additive homomorphic cryptosystems in Sect. 4.5.3.

4.2.2 Blockchains and Smart Contracts

For the proposed protocol, blockchains are used as a distributed, public, and permanent platform for deciding on an optimum tariff and storing this decision reproducibly. A blockchain as a public ledger for coin-based transactions is originally proposed in [5] for *Bitcoin*. Since then, a wide range of applications have been established that are based on that technology (e.g., [6, 21–23]). Some approaches improve the principles of Bitcoin in terms of privacy [21], while others propose a turing-complete blockchain for arbitrary calculations [6, 23].

In [22], the authors propose a protocol that uses a blockchain as an access-control manager that does not require a third party. This allows to reduce trust into a single entity, similarly to the approach proposed in this paper.

In [6], the author presents a turing-complete blockchain that can be used to build decentralized applications. All inputs and states are public in this approach, which limits privacy. In [23], the authors propose a similar concept that preserves user privacy by the use of zero-knowledge proofs.

4.2.3 *Other Related Work*

In our proposed approach, we make use of oblivious transfer [10, 24, 25]. It allows one of two parties to query one of an arbitrary number of items from the second party without revealing (1) which item has been requested; and (2) any of the other items. A detailed description of the use within our protocol is provided in Sect. 4.3.2.

From a use case point of view, apart from oblivious transfer, related work includes literature on tariff decisions, load profile matching and forecasting as well as demand-response and demand-side-management. For the latter two, Palensky and Dietrich [26] provide an overview. In general, privacy concerns and communication overhead on the smart meter side are seldomly addressed in protocols for demand side management and tariff-matching. We focus on these privacy and communication overhead aspects by providing some examples below.

Caron and Kesidis [27] describe an approach where customers share load profile information so that the utility can achieve a smoother aggregate load profile. This is also possible with the approach proposed in our paper. However, the load profiles are not revealed to the utility, thus preserving the privacy of the customers.

Similarly, Shao et al. [28] attempt to reduce peak loads by incentivizing customers to shift time of use of electrical devices. This behavior can also be triggered when applying our approach to provide suitable template load profiles, albeit not in real-time. Again, the load profiles do not need to be shared with the utility as opposed to the approach by Shao et al.

Ramchurn et al. [29] follow a game-theoretic approach with customer incentives for reducing peak loads. They use a decentralized protocol, as opposed to our as well as to Caron's and Kesidis's [27] and Shao et al.'s [28] approach. The approach by Ramchurn et al. defers certain loads with defined probability imposing constraints on the customer's choices. In contrast, our approach gives the customer full authority on tariff decisions.

Another game-theoretic approach for shifting energy consumption is proposed by Mohsenian-Rad et al. [30]. In this setting, the smart meters of the users interact in order to minimize overall energy consumption. Their approach requires peer-to-peer communication with transmission overhead depending on the number of smart meters involved. Conversely, our approach does not require any peer-to-peer communication whatsoever.

4.3 Preliminaries

This section introduces the preliminaries for the proposed protocol that uses blockchains and smart contracts for the matching process. First, the previously proposed protocol [4] that uses a dedicated third party for the matching process is presented. Second, the concepts of blockchains and smart contracts are introduced as a decentralized and trust-less alternative for the original protocol.

Table 4.1 Overview of notation used in this paper

U	Utility provider
SM	Smart meter
TP	Third party
$\mathbf{F}$	Load profile forecast
$\tilde{\mathbf{F}}$	Embedded load profile forecast
$\mathbf{T}$	Template load profile
$\tilde{\mathbf{T}}$	Embedded template load profile
T	Tariff
$\mathscr{U}$	Set of utilities
$\mathscr{L}_u$	Set of tariffs from utility u
(u^*, l^*)	Index of best-matching utility/tariff
$E(\cdot), D(\cdot)$	Encryption and decryption functions
$r \leftarrow_\$ \mathbb{Z}$	Sampling of a random number from $\mathbb{Z}$

The notation used throughout this paper is summarized in Table 4.1. $\mathbf{F}$ denotes a load profile forecast that is used by the consumer to represent the load pattern and $\mathbf{T}$ denotes a template load profile that is linked to a tariff from the utility providers. The embedding is explained in the next section and is denoted as $\tilde{\mathbf{F}}$ and $\tilde{\mathbf{T}}$, respectively.

4.3.1 Requirements

The objective of this paper is to rely on a fully decentralized and trust-less architecture for tariff-matching. For the matching process and the performed calculations, we define the following requirements:

1. **Transparency.** All calculations should be transparent to both, the utility providers and the customers. This prevents an adversary from forging matching results (i.e., a binding to committed values) and it also prevents the utility providers from later changing their template load profiles retroactively.
2. **Verifiability.** Calculations and the results, i.e., the optimum matching should be verifiable by all participants. This allows the customer and the utility providers to verify at any time that the found tariff has indeed a minimum distance with the submitted parameters.
3. **Reliability.** The matching should not rely on a single party. All participants should be able to perform the matching and therefore increase the fault-tolerance of this approach by decentralizing this crucial step in the protocol. In the original paper [4], the protocol requires a dedicated third party that acts as a broker for the matching.
4. **Privacy.** While fulfilling the first three requirements, the calculation should still be privacy-preserving in accordance to the following privacy definition.

Privacy Definition: A protocol is privacy-preserving if none of the participants learns more than a particular, predefined function of the input data [31]. The proposed protocol is privacy-preserving, if none of the participating entities has access to the original load profiles and only the minimum distance between the customers' load profile forecast and the template load profiles of the utility provider is revealed. At the end of the protocol, the smart meter only learns the tariff associated to a template load profile that has the minimum distance to the load profile forecast.

4.3.2 *Profile Matching Protocol*

This section is a summary of our previous protocol [4] which allows one smart meter to find an optimum tariff based on its load profile forecast. A number of template load profiles corresponding to tariffs from different utilities are used for this search. Throughout the process, none of the involved parties has access to the others' data. The process is therefore privacy-preserving. Note that the following protocol still makes use of the third party for finding the optimum match. It will later be shown how to replace this with a blockchain and smart contracts, respectively.

Figure 4.1 illustrates the different steps of our protocol which are described in detail below. The protocol is split into three phases, namely *initialization*, *matching*, and *oblivious transfer*.

4.3.2.1 Initialization

Let the set of participating utilities be denoted as $\mathscr{U}$. Each utility $\mathbf{U}_u$, $u \in \{1, \ldots, |\mathscr{U}|\}$ (cf. Fig. 4.1, right) has a list of tariffs $T_{u,l}$ with corresponding template load profiles $\mathbf{L}_{u,l}$ (denoted as set $\mathscr{L}_u$ in Fig. 4.1 where utilities can have different numbers of load profiles $l \in \{1, \ldots, |\mathscr{L}_u|\}$). For example, utility $\mathbf{U}_1$ has a standard tariff $T_{1,1}$ for day workers and a night-owl tariff $T_{1,2}$ for night workers. The corresponding template load profiles $L_{1,1}$ and $L_{1,2}$ show peak loads at different times of the day opposite to the respective working hours.

Template load profiles allow each utility to control demand and response in a fine-grained manner, e.g., by rewarding customers with atypical load profiles so that peak loads are avoided. Although the tariffs need to be known to the customers for billing and transparency, the template load profiles are considered to be private to the respective utility. Therefore, no details are shared with competing utilities. In practice, template load profiles and tariffs may be significantly more complex than the example described above, with privacy on the utility side being a much more pressing issue.

In order to keep the template load profiles private, each utility calculates an embedding $\tilde{\mathbf{L}}_{u,l} \in \{0, 1\}^m$ for each of its original template load profiles $\mathbf{L}_{u,l} \in \mathbb{R}^k$ as the first step of the *initialization* phase:

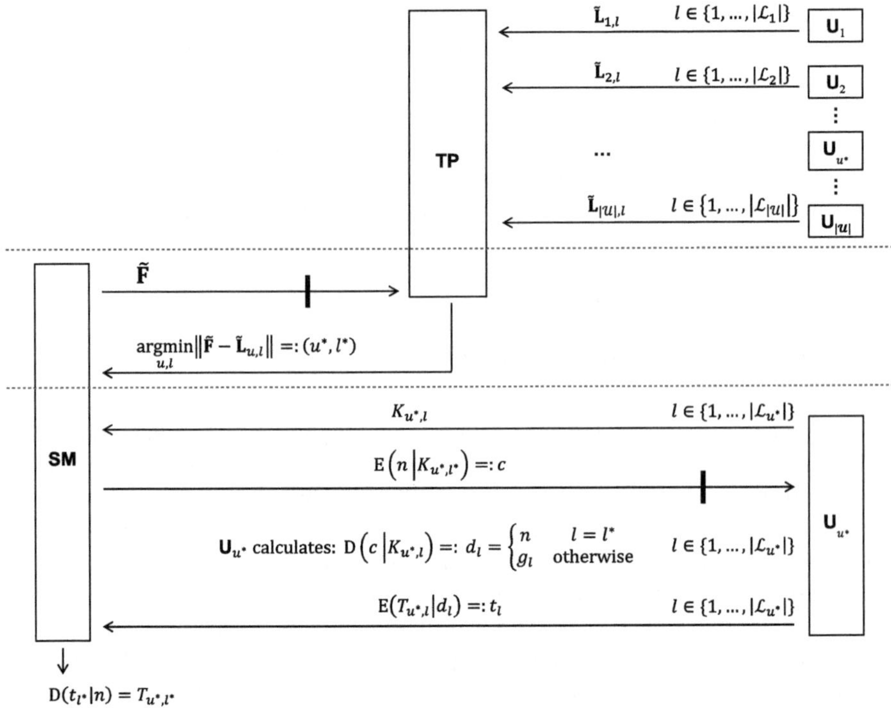

Fig. 4.1 Overview of the proposed protocol. All participating utilities $\mathbf{U}_u \in \mathcal{U}$ provide embedded template load profiles $\tilde{\mathbf{L}}_u$ corresponding to their tariffs $T_{u,l}$ to a third party **TP**. A smart meter **SM** sends its embedded load profile forecast $\tilde{\mathbf{F}}$ to this third party which returns the best-matching utility index u^* and its tariff index l^*. Subsequently, the smart meter can request information about the best-matching tariff T_{u^*,l^*} from the best-matching utility $\mathbf{U}_{u^*}$ through oblivious transfer. *Black vertical bars* indicate rate limiters

$$\tilde{\mathbf{L}}_{u,l} = \left\lceil \frac{\mathbf{A} \cdot \mathbf{L}_{u,l} + \mathbf{W}}{\Delta} \right\rceil \mod 2 \tag{4.1}$$

$\mathbf{A}$ is a random $m \times k$ matrix with i.i.d. Gaussian elements with mean 0 and variance σ^2, and $\mathbf{W}$ is a random m-dimensional vector with i.i.d. uniform elements in the range $[0, \Delta]$. Δ is both, a quantization and a security parameter, and described in detail in [9, 14].

This embedding does not allow a potential attacker to reconstruct the original load profile, but preserves the distance between load profiles within a very small margin of error as described below. This enables comparisons of load profiles without the need to handle the respective original, private data.

The second step of the *initialization* phase of our protocol requires each utility to send all of its calculated embeddings $\tilde{\mathbf{L}}_{u,l}$ to a third party, denoted as **TP**. The need for this third party will become clear in the subsequent *matching* phase.

4.3.2.2 Matching

In this phase, a smart meter, denoted as **SM**, first creates a load profile forecast $\mathbf{F}$. It can either be based on past load profiles, e.g., of the current day or week, or on user input, e.g., prospective changes in work schedules. Similar to each utility in the preceding *initialization* phase, the smart meter first calculates an embedding of $\mathbf{F}$, denoted as $\tilde{\mathbf{F}}$. This way, the smart meter does not need to disclose its original load profile which may reveal sensitive information about the user.

As a second step, the embedding is sent to the third party, like the embeddings from the utilities in the previous phase. As a third step, the third party finds the best match for the load profile forecast out of the list of template load profiles from all utilities through the received embeddings. More precisely, it finds the template load profile with the smallest normalized Hamming distance to the forecast and outputs the template load profile index l^* as well as the corresponding utility index u^*, i.e.,

$$(u^*, l^*) = \underset{u,l}{\operatorname{argmin}} \, ||\tilde{\mathbf{F}} - \tilde{\mathbf{L}}_{u,l}||_1. \tag{4.2}$$

This is possible due to the distance-preserving property of the embeddings (as described in more detail in [9]), where the Euclidean distance of the original data vectors is proportional to the normalized Hamming distance of the embedded vectors with a configurable small error ε, i.e.,

$$||\tilde{\mathbf{F}} - \tilde{\mathbf{L}}_{u,l}||_1 \sim ||\mathbf{F} - \mathbf{L}_{u,l}||_2 + \varepsilon. \tag{4.3}$$

As a consequence, the probability that the best match in the original space and the best match in the embedded space coincide is close to one:

$$\Pr\left[\underset{u,l}{\operatorname{argmin}} \, ||\tilde{\mathbf{F}} - \tilde{\mathbf{L}}_{u,l}||_1 = \underset{u,l}{\operatorname{argmin}} \, ||\mathbf{F} - \mathbf{L}_{u,l}||_2\right] = 1 - \delta. \tag{4.4}$$

This probability is referred to as matching accuracy. The smaller ε is, the higher the accuracy is.

The result (u^*, l^*) of the matching operation is a pair of indices identifying the utility $\mathbf{U}_{u^*}$ of the best match and its tariff T_{l^*}. However, no information about any load profile is revealed. The tuple (u^*, l^*) is transmitted to the smart meter as a fourth and final step, allowing the smart meter to fetch the tariff information from the utility $\mathbf{U}_{u^*}$ directly in the next phase in order not to disclose the actual tariff T_{l^*} associated with the index l^*.

The third party needs to be involved in the calculation above since neither the smart meter nor any of the utilities can be completely trusted to correctly perform calculations on the data. In addition, malicious parties are considered, i.e., they could manipulate their own input to bias the result in their favor or they could use multiple different inputs to derive additional information about the other parties' data. This is avoided by the use of an independent third party with a rate limiter (vertical black bars in Fig. 4.1) which prevents bulk-probing from the other parties.

The third party can be thought of as a neutral party which performs only computations, e.g., a proxy of the Council of European Energy Regulators (CEER) which strives for a fair tariff market and competition. However, since any party may be distrusted, including the third party, the latter is not allowed to perform calculations on the actual data, but on embeddings only. This is why the latter need to be calculated by the other parties. This way, the third party is not able to access the original load profiles of either the smart meter or the utilities.

Note that the third party may collect statistics on the matching results (i.e., the indices (u^*, l^*)) of all smart meters. However, this can be rendered futile if the utilities regularly shuffle their template load profiles' indices in the *initialization* phase, e.g. each day. For example, $(u^*, l^*) = (1, 1)$ means a standard tariff and $(u^*, l^*) = (1, 2)$ a night-owl tariff, respectively, on one day, and the other way around, i.e., $(u^*, l^*) = (1, 1)$ means a night-owl tariff and $(u^*, l^*) = (1, 2)$ a standard tariff, respectively, on the next day.

Note that the third party could be omitted when using verifiable computing [32, 33]. However, this would induce substantial overhead which is critical on a device with limited capabilities, such as smart meters. In the subsequent *oblivious transfer* phase, the third party is not involved at all and hence never has access to the tariff T_{u^*,l^*} itself. Therefore, the third party only needs to be trusted to perform the correct calculation in the *matching* phase. This remaining level of trust will be replaced by a decentralized trust-less architecture for the blockchain-based protocol.

4.3.2.3 Oblivious Transfer

In the last phase of the embedding-based protocol, the smart meter sends a query to the utility $\mathbf{U}_{u^*}$ in order to obtain the best-matching tariff T_{l^*} based on its index l^*. The third party is not involved in this transaction and does therefore not obtain any information about the tariff itself apart from its index.

At this stage, the customer has not yet made the decision whether or not to switch to the best-matching tariff—they still need the tariff information for this. Thus, on the one hand, the smart meter must not disclose l^* to the utility since it would allow the utility to deduce information about the original load profile, even when the customer chooses not to switch to the matching tariff. On the other hand, the utility does not want to disclose all tariffs, some of which may exclusively be available to certain customers or groups. It only wants to disclose the tariff corresponding to the index l^*, but without being allowed to know this very index.

A solution for this is oblivious transfer [10, 24, 25]. It allows the smart meter to retrieve a tariff l^* from a vector of tariffs $T_{u^*,l}$, without the query (index) being known to the utility $\mathbf{U}_{u^*}$ and without any other tariffs being disclosed to the smart meter apart from T_{u,l^*}. In our use case, communication with $\mathbf{U}_{u^*}$ yields the tariff T_{u^*,l^*}.

All encryptions and decryptions are performed with a public-private key cryptosystem. For low overhead and smaller key sizes elliptic curve cryptography can be used with a recommended key size of 256 bit [34, 35].

The steps of the *oblivious transfer* phase can be summarized as follows: Initially, i.e., before any other communication, the utility $\mathbf{U}_{u^*}$ sends one public key $K_{u^*,l}$ per template load profile $\mathbf{L}_{u^*,l}$ to the smart meter. The number of template load profiles is identical to the number of tariffs as described above. Thus, in total $|\mathscr{L}_{u^*}|$ keys are sent.

Secondly, the smart meter generates a nonce n which is encrypted with the (l^*)th key. The encrypted nonce, c, is then sent to the utility. This step requires a rate limitation (e.g., one query per day) on the utility's side since the smart meter could otherwise query all available tariffs by iterating through the available indices. The rate limit has to be chosen such that the maximum number of allowed queries is lower than or equal to the average frequency at which utilities update their tariffs. If, for instance, utilities update their template load profiles daily, a rate limit of one query per day is sufficient.

Thirdly, the utility decrypts c with all of its private keys, yielding decryptions d_l. For the key with the index sent by the smart meter, the decryption yields the nonce n, while, for all other keys, the decryption result is a garbage value g_l. For the utility, however, these are indistinguishable from the nonce. Thus, the utility cannot find out the index l^*.

Fourthly, the utility encrypts all tariffs $T_{u^*,l}$ with the decryption results d_l as the public keys, i.e., the nonce known to the smart meter for the index l^* and garbage values g_l for the others. The encrypted tariffs, t_l, are then sent to the smart meter which decrypts the (l^*)th encrypted tariff with the previously generated nonce as the private key. This yields the desired tariff T_{u^*,l^*}. The other tariffs cannot be decrypted since the encryption and decryption keys do not match, thus do not leak any information to the smart meter.

4.3.3 *Blockchains and Smart Contracts*

This section introduces blockchains and the concept of smart contracts. Furthermore, an exemplary smart contract is presented, the application of a decentralized, trust-less architecture is motivated and privacy issues are discussed.

4.3.3.1 Overview

After having originally been proposed in [5], blockchains are gaining an increased adaption in many fields, e.g., [22, 23]. A blockchain is a trust-less and fully decentralized peer-to-peer system that is designed to hold immutable information once data is committed to the chain. Generally, a blockchain can therefore be described as a distributed, immutable database.

In the originally proposed Bitcoin protocol from [5] the blockchain is used to keep track of *coins*, i.e., a public list of how much coins are owned by each peer. Therefore, each block contains sender and receiver information, as well as the amount of coins to be transferred. This is called a *transaction* and—once confirmed by the peers—appends a new block to the chain that also includes the hash of the previous block and is therefore permanently linked to a series of previous transactions.

The public list of chained blocks can be verified by all peers in the network by checking the integrity of the new block and the correct calculation of the hash, respectively. While the generation of valid blocks consumes a considerable amount of computing power, this is also referred to as the *proof of work*. In order to prevent the tampering with already created blocks, all peers in the network agree on the longest valid chain. A chain is valid if it is verified by other peers. Accordingly, a block and transaction can be considered valid if it is followed by a sufficient amount of other valid blocks. Therefore, the blockchains is trust-less if at least half of the computing power for the proof of work is spent by honest peers [5].

In order to prevent the tampering with already created blocks, all peers in the network agree on the longest valid chain. A chain is valid if it is verified by other peers. Accordingly, a block and transaction can be considered valid if it is followed by a sufficient amount of other valid blocks. Therefore, the blockchains is trust-less if at least half of the computing power for the proof of work is spent by honest peers [5].

Note that at all times all states are publicly available and can be verified by all peers by simply checking the hashes from the very first block (also referred to as the genesis block) up to the last block. Peers in the network are identified by a private-public key pair (identifier or address).

4.3.3.2 Smart Contracts

The Bitcoin protocol is designed for a particular purpose. However, blockchains can be used to store arbitrary information. Recently, the application of blockchains for smart contracts has been proposed [6, 23, 36]. Ethereum,[2] for instance, is a platform for executing smart contracts on a turing-complete virtual machine, the Ethereum Virtual Machine (EVM), where computing results are stored in a public blockchain. Similarly to the Bitcoin protocol, in Ethereum, a peer-to-peer network stores a decentralized ledger with state information. These states, however, include details about executed smart contracts. A smart contract is a program that can send and receive messages and execute some logic. For executing a contract, i.e., creating a new block, the miner is paid with *Ether*. There are a number of newly developed program languages that compile to EVM bytecode, such as Solidity[3] that resembles a JavaScript-like syntax.

Smart contracts can be deployed by every peer in the network. Once committed to the blockchain, a smart contract is identified by a unique address and can be called by other peers. The code of a smart contracts needs to be public as well,

---

[2]https://www.ethereum.org/.

[3]https://solidity.readthedocs.io/en/develop/.

Table 4.2 Overview of notation used in algorithms in this paper

ID	Data type holding an address in the blockchain
Hash	Data type holding a hash (e.g., SHA-2)
HashMap<U,T>	Data type mapping elements of type U to elements of type T
$\tilde{\text{E}}$	Data type holding an embedding
$\tilde{\text{E}}$[]	Data type holding an array of embeddings
$\oslash$	Value representing null or undefined

since all peers in the network must be capable of verifying the computation results. Smart contracts therefore expose one or more methods and hold state information in internal variables. As a simple example which is commonly used to demonstrate a smart contract (compare to, e.g., [23]) consider a rock-paper-scissors game as in Algorithm 1. Here, a smart contract is created that accepts an input from two players ("rock," "paper," or "scissors") and determines a winner. The notation for algorithms in this paper is summarized in Table 4.2. Assume this algorithm is deployed in the blockchain and is assigned a unique identifier or address. Players willing to bet can use this address and call the `commit` method in order to send their bet.

The first parameter of type ID is passed automatically as an argument to all methods and refers to the address of the caller. This is a concept that is also found in real-world implementations. For the algorithms in this paper this is made explicit for clarity. The address is an alpha-numeric value of approximately 40 characters, depending on the concrete protocol that is derived from the public key of a peer.

In order to prevent any player from learning the other players' value, a computationally hiding commitment scheme is used (see [37]). Instead of only sending a string *bet*, player i, $i \in \{1, 2\}$, sends the hash of the string and a random number $r_i \leftarrow_{\$} \mathbb{Z}$, i.e., $h_i := H(\text{bet}_i|r_i)$ with $H(\cdot)$ being a collision-resistant universal hash function $H : \{0, 1\}^x \rightarrow \{0, 1\}^y$, $x, y \in \mathbb{Z}^+$, with, e.g., $y = 256$, and $\text{bet}_i|r_i$ denotes the concatenation of the bet and the random number. Due to the one-way property of the hash function and the random number which is only known to the player, the public information in the blockchain does not reveal the bet, but only the hash. Once all players sent their commitments, the game can be evaluated. To do so, the players open their commitment by sending (bet_i, r_i) and the smart contract verifies whether $H(\text{bet}_i|r_i) = h_i$. If the hash of the commitment and the hash of the values bet_i and r_i are equal, the commitment is opened. Otherwise, the commitment is invalid and the winner cannot be determined until the commitment is opened correctly. The message flows for player i and the smart contract are shown in Fig. 4.2. Note that there also exist information-theoretically secure binding schemes such as Pedersen commitments [38], which are out of scope for this work.

While such commitment schemes prevent other players from learning the bets during the game, they do not provide forward secrecy once the commitments are opened and the game is finished. This would, for instance, allow an adversary to

Algorithm 1: Smart contract for playing a rock-paper-scissors game. Both players send a commitment for their bet and once all bets are received, the players open their commitment and the winner can be determined by calling the `evaluate` method

```
ID player1 = ⊘;
ID player2 = ⊘;
Hash hash1 = ⊘;
Hash hash2 = ⊘;
string bet1 = ⊘;
string bet2 = ⊘;

begin commit (ID sender, Hash hash)
    if player1 == ⊘ then
        player1 = sender;
        hash1 = hash;
    else if player2 == ⊘ then
        player2 = sender;
        hash2 = hash;
    else
        return "wait for next game";
    end
end

begin open (ID sender, string bet, int r)
    if sender == player1 ∧ H(bet|r) == hash1 then
        bet1 = bet;
    else if sender == player2 ∧ H(bet|r) == hash2 then
        bet2 = bet;
    else
        return "invalid commitment";
    end
end

begin evaluate (ID sender)
    if ¬ (bet1 == ⊘ ∨ bet2 == ⊘) then
        if bet1 == "rock" ∧ bet2 == "scissors" then
            return player1 + " wins";
        else if bet1 == "paper" ∧ bet2 == "scissors" then
            return player2 + " wins";
        else
            ... /* test for other cases */
        end
    else
        return "commit bets and open commitment first";
    end
end
```

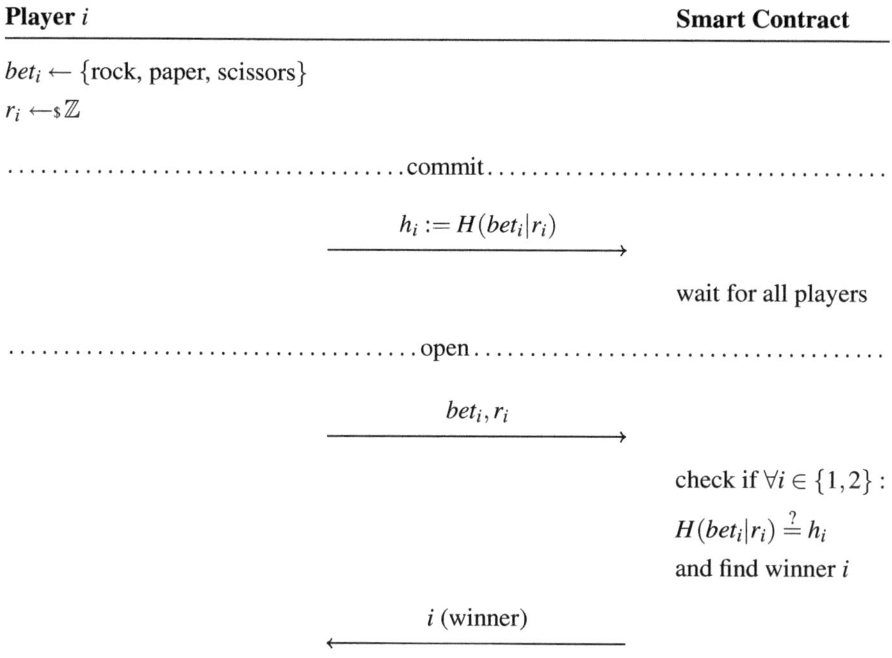

Fig. 4.2 Message flows for the rock-paper-scissors game for player i, $i \in \{1, 2\}$, and the smart contract with a computationally hiding commitment scheme. First, each player chooses a bet and a random number, where $r \leftarrow_{\$} \mathbb{Z}$ denotes the drawing of the random number. Both values are hashed and committed. Second, once all players sent their values, the commitments are opened. If the values are correct, the winner is determined

learn the bets from a player over time.[4] A more advanced approach is to use zero-knowledge proofs, e.g., as presented in [23], where the authors present a blockchain model for smart contracts that allows for private parts that are never revealed publicly.

4.3.3.3 Summary

Generally, blockchains such as Bitcoin and blockchain-based smart contract implementations such as Ethereum are not privacy-preserving in the sense that the information remains undisclosed. In contrast, all the information committed to the blockchain is publicly available to all peers. Usually, the ownership of coins and other data is claimed by a private key that has been randomly generated once and

[4]A player could change address for every game providing a means of pseudonymity for this use case. However, for the tariff matching, as shown later, we need to reveal the actual "players" at some point for the oblivious transfer which requires more sophisticated approaches.

does not allow to link to any individual. However, this kind of pseudonymity is not a strong privacy guarantee, as sender, receiver, and the amount of data as well as state information are public. In order to privately process data in a blockchain, there are recent trends that strengthen the privacy, e.g., [21, 23].

In summary, blockchains offer the ability to provide a decentralized database that does not require trusting other peers with certain limits as discussed in Sect. 4.3.3.1. A consensus on the state of the database is found by a set of decentralized executed rules and proof-of-work mechanisms. Blockchains are therefore particularly suitable for minimizing or removing the role of a trusted or semi-trusted third party, while at the same time providing full transparency to all participants. However, in order to achieve privacy, blockchains need more sophisticated approaches, e.g., building on commitments or zero-knowledge proofs. It is later shown that, for preserving forward-secrecy in the proposed protocol, embeddings and commitments are not sufficient and stronger privacy guarantees as proposed in [23] are required.

4.3.4 Assumptions

For the following protocol description, we define a number of properties for our blockchain-based profile matching protocol. First, a preliminary protocol is introduced that fulfills already three out of our four initial requirements: (1) Transparency; (2) Verifiability; and (3) Reliability. The fourth requirement, privacy, is only partly fulfilled. While the template load profiles and the load profile forecasts are embedded and do not allow for immediate recovery of the original data, all this information is publicly and permanently available and stored in the blockchain, i.e., forward secrecy is not fulfilled. An adversary could query the blockchain for previous requests and use this information to brute-force load profiles. The feasibility of brute-forcing embedded data is briefly discussed in [4].

In order to fulfill our privacy requirement regarding forward-secrecy, a way needs to be established that allows to calculate the matching in the blockchain while at the same time not revealing the inputs.

For the preliminary protocol, we assume a blockchain that allows to execute smart contracts (e.g., Ethereum). Therefore, all parties, i.e., **SM**, $\mathbf{U}_u$ with $u \in \{1, \ldots, |\mathcal{U}|\}$ are peers in the blockchain and are thus able to execute smart contracts. For the improved protocol we additionally assume the blockchain to be capable of privacy-preserving calculations (e.g., HAWK [23]).

4.4 Protocol Description

Figure 4.3 illustrates the different steps of our adaption of the original protocol for blockchains. The protocol can again be split into three phases, namely *initialization*, *matching*, and *oblivious transfer*, with the *matching* phase being further split into a *commitment* and an *opening* phase (separated by red dotted lines). In the

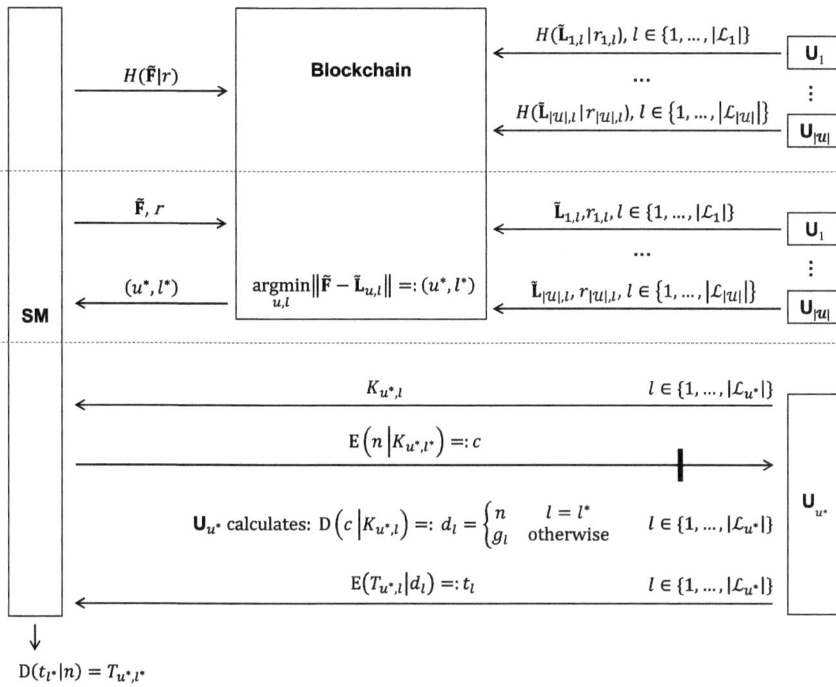

Fig. 4.3 Overview of the proposed protocol. After the *initialization* phase (not shown), all participating parties commit their load profiles to the block chain as the first part of the *matching* phase. In the second part, all parties open their commitments publicly in the blockchain. At the end of the *matching* phase, the established smart contract returns the best-matching utility index u^* and its tariff index l^*. The smart meter can then request information about the best-matching tariff through oblivious transfer as proposed in [4]

initialization phase, the same embedding approach as in the previous protocol is applied. In this phase, load curves are transformed into an embedding that does not allow to retrieve the original load profile, but preserves the ability to calculate distances. In contrast to the original protocol, the matching phase is handled within a public blockchain instead of a dedicated third party. The *oblivious transfer* phase remains the same as in the previous protocol and is therefore not discussed in detail in this section.

4.4.1 Initialization

Each utility $\mathbf{U}_u$, $u \in \{1, \ldots, |\mathscr{U}|\}$ (cf. Fig. 4.3, right) has a list $\mathscr{L}_u$ of tariffs $T_{u,l}$ with corresponding template load profiles $\mathbf{L}_{u,l}$. As in [4], utilities can have different numbers of load profiles $l \in \{1, \ldots, |\mathscr{L}_u|\}$).

For each template load profile $\mathbf{L}_{u,l}$, its embedding $\tilde{\mathbf{L}}_{u,l} \in \{0, 1\}^m$ is calculated by $\mathbf{U}_u$ as specified by Eq. (4.1). This way, in the subsequent *matching* phase, utilities do not need to submit the plain template load profiles, but only an embedding for privacy reasons. Similarly, **SM** calculates the embedding $\tilde{\mathbf{F}}$ of its load profile forecast $\mathbf{F}$ and only sends $\tilde{\mathbf{F}}$ in the *matching* phase.

4.4.2 Matching

In contrast to the original paper, the *matching* phase is completely redesigned, fully decentralized, and trust-less based on a blockchain. Using a blockchain and a smart contract for determining the distance poses some advantages over a dedicated entity **TP**, which corresponds to the initially stated requirements:

1. **Transparency.** Once a template load profile $\tilde{\mathbf{L}}_{u,l}$ or a load profile forecast $\tilde{\mathbf{F}}$ is written to the blockchain, this information is public. Every peer can view the data and the data cannot be altered by anyone after being encoded in the chain.
2. **Verifiability.** The calculation performed by the smart contract, i.e., the result of the calculation which in this case is the optimum match, is written immutably to the blockchain as well. Therefore, every peer can later verify the chain of blocks and whether the calculated result is actually the best match. In particular, this also holds for the customer.
3. **Reliability.** Smart contracts that are based on blockchains are decentralized by design. There is no single entity that is responsible for performing the calculations, but every peer can calculate and verify results. This makes blockchains more reliable, compared to centralized architectures.

The *matching* phase that only needs to compute the hamming distance of the embedded load forecast and the embedded template load profile [as shown in Eq. (4.2)] is written as a smart contract. This smart contract is created by the smart meter **SM** which prepared an embedded load profile forecast $\tilde{\mathbf{F}}$ in the *initialization* phase and now wants to find an optimum tariff. The address in the blockchain of the smart meter and the load profile forecast are the initialization parameters for this smart contract. The contract is then bound to this particular smart meter. The binding assures that no other peer can trigger the evaluation method, i.e., the initial creator remains in control of the contract.

Once this smart contract is created, utility providers $\mathbf{U}_u$, $u \in \{1, \ldots, |\mathscr{U}|\}$ can commit their embedded template load profiles $\tilde{\mathbf{L}}_{u,l}$, with $l \in \{1, \ldots, |\mathscr{L}_u|\}$ by sending $H(\tilde{\mathbf{L}}_{u,l}|r_{u,l})$, where $r_{u,l}$ is the random number for the commitment as described in Sect. 4.3.3.2. The collection of load profiles can be realized by providing a public method that accepts an array of hashed template load profiles as a commitment from each utility in the *commitment* part (top part of Fig. 4.3). As soon as a smart contract is established, the commitment of the load profile forecast from the customer as well as any committed template load profiles from the utility providers cannot be changed or overwritten in this particular instance. However, if a new smart contract is created, both, customer and utility providers can commit their load profiles again and the protocol starts over. The smart contract is open and accepts the commitments from the utility providers as long as the commitments are not opened.

If the smart meter received enough tariff options or after a certain amount of time has passed, **SM** and all $\mathbf{U}_u$ can trigger the opening method, thereby starting the *opening* part (middle part of Fig. 4.3). If the commitments are opened correctly, the actual embedded load profiles are stored in the smart contract, **SM** triggers the evaluation method and the best-matching tariff (u^*, l^*) is returned. This also closes the current instance of the smart contract. Note that the index u^* of the utility corresponds to the **ID** or address of the utility in the blockchain.

4.4.3 *Oblivious Transfer*

The *oblivious transfer* phase does not deviate from the originally proposed protocol and is described in Sect. 4.3.2. The blockchain is only used for finding the minimum distance to the utility/tariff that matches best, i.e., (u^*, l^*). The smart meter then initiates the oblivious transfer in order to receive the desired tariff T_{u^*,l^*}.

4.5 Evaluation

In this section, we evaluate privacy-preserving load profile matching. For the evaluation we introduce an implementation for the smart contract that is independent of concrete programming languages or platforms. We then discuss the privacy and security properties of the blockchain-based approach and finally, the proposed approach is compared to related work.

4.5.1 *Implementation*

The following algorithms, Algorithm 2 and Algorithm 3, show how to implement a smart contract that performs tariff matching. Note that this is an implementation to show the concept and that this code still has some practical limitations as discussed later.

While the following algorithm is independent of concrete smart contract programming languages (e.g., Solidity, Hawk), it follows common design principles [37] and can be easily turned into a program in a specific language. The smart contract consists of five global variables and five methods that implement the creation and evaluation of the smart contract as well as the ability to receive and

Algorithm 2: This first portion of the smart contract for matching a load profile forecast to template load profiles shows the creation, commitment and open methods for both, the smart meter and the utilities. The parameters of type **ID** always refer to the caller. A smart meter initiates a matching by creating a smart contract with a commitment for its load profile forecast. Utilities then commit their template load profiles. The smart meter and the utilities can open the commitments before determining the best-matching tariff.

```
ID owner = ∅;
Hash loadforecasthash;
HashMap<ID, Hash[]>templatehashes = new HashMap<>();
Ẽ forecast = ∅;
HashMap<ID, Ẽ[]>templates = new HashMap<>();

begin create (ID sender, Hash loadforecast)
    owner = sender;
    loadforecasthash = loadforecast;
end

begin commit (ID sender, Hash[] loadprofiles)
    if ¬ templatehashes.contains(sender) then
        templatehashes.add(sender, loadprofiles);
    else
        return "load profile commitments already sent";
    end
end

begin smopen (ID sender, Ẽ loadforecast, int r)
    if sender == owner ∧ H(loadforecast|r) == loadforecasthash then
        forecast = loadforecast;
    else
        return "invalid commitment";
    end
end

begin uopen (ID sender, Ẽ[] loadprofiles, int[] r)
    if templatehashes.contains(sender) then
        int index = 0;
        foreach Hash embeddinghash in templatehashes.getValues(sender) do
            if ¬ (embeddinghash == H(loadprofiles[index]|r[index])) then
                return "invalid commitment";
            end
            index ++;
        end
        templates.add(sender, loadprofiles);
    end
end
```

Algorithm 3: This second portion of the smart contract for matching a load profile forecast to template load profiles shows the evaluation to find the minimum distance. If all commitments have been opened, the creator of the smart contract can evaluate the template load profiles and determine the best-matching tariff and the corresponding utility offering that tariff

```
begin evaluate (ID sender)
    if sender == owner ∧ ¬ (forecast == ⊘) ∧ ¬ templates.empty() then
        double bestdistance = ∞;
        ID bestutility = ⊘;
        int besttariffindex = ⊘;
        foreach ID utility in templates do
            int index = 0;
            foreach Ẽ embedding in templates.getValues(utility) do
                double distance = ‖embedding − forecast‖;
                if distance < bestdistance then
                    bestdistance = distance;
                    bestutility = utility;
                    besttariffindex = index;
                end
                index ++;
            end
        end
        return (bestutility, besttariffindex);
    end
    return "not allowed";
end
```

verify commitments. The smart contract uses basic data types such as double precision floating-point types (*double*) and integer types (*int*), more complex data types such as *HashMap* and Arrays (denoted by square brackets), specific data types such as **ID** (storing an address of a node in the blockchain), $\tilde{\mathbf{E}}$ (storing an embedding, which is discussed in detail later), and **Hash** (storing a SHA-2 hash). The principle of this smart contract is similar to the implementation presented in Algorithm 1 for the rock-paper-scissors game. All parties first send hashed values as a commitment and then open their commitments, which is verified in the smart contract by comparing the hashes. Instead of determining a winner as in the game, the optimum tariff based on the minimum distance is found.

In the following, the methods for the smart contract are described in detail:

- `create`. This method is called upon the initial creation of the smart contract. The method expects two arguments, *sender* and a hashed value *loadforecast*. The first is the creator of the smart contract, i.e., the smart meter that wants to perform a tariff matching. The latter is a commitment for an embedding of the load profile forecast. Both values are used for initializing the smart contract. The commitment is calculated by the smart meter by hashing the embedded load profile forecast and a random number.

- `commit`. Once the smart contract is initialized, it can be found by utilities that want to offer a tariff. These utilities then call the *commit* method which expects two arguments, *sender*, which is the address of the utility, and *loadprofiles*, which is an array of hashed embedded template load profiles. Utilities can only once send an array of an arbitrary number of hashed template load profiles. This assures that utilities cannot change or revoke committed data for this instance of the smart contract. The commitment is calculated by the utilities by hashing the embedded template load profiles and a random number for each template load profile.
- `smopen`. The smart meter can open its commitment by sending the embedded load profile forecast and the random number that was used for calculating the hash. The smart contract verifies if the commitment is valid and stores the embedded load profile forecast for the evaluation.
- `uopen`. Similarly to the smart meter, utilities can open their commitments by sending the embedded template load profiles and the random numbers that were used for calculating the hashes. The smart contract verifies if the commitment is valid and stores the template load profiles for the evaluation.
- `evaluate`. After a certain amount of time or whenever the smart meter feels that enough utilities have committed and opened their template load profiles, the smart meter can call the *evaluate* method. This method can only be called if both, the smart meter and the utilities have successfully opened their commitments and it can only be called by the owner of the smart contract. The method performs the actual matching [i.e., finding the minimum distance in the embedded dimension as shown in Eq. (4.2)] and returns both, the address of the utility with the best-matching tariff and the index of that tariff (u^*, l^*). This terminates the execution of the smart contract and the smart meter uses this information to run the *oblivious transfer* phase outside of the blockchain and the smart contract.

For an actual implementation of such a smart contract, there are a few aspects to be considered. First, the evaluation of the original protocol [4] has shown that an embedding dimension of $m = 8192$ is needed, which consequently leads to 8192 bit numbers that need to be processed for the matching. The EVM, for instance, has maximum word size of 256 bit which would require additional steps to process numbers of that size.

Second, the above smart contract fully relies on the privacy features of embeddings. After the smart meter opened its committed load profile forecast or a utility opened its committed template load profile, respectively, this information is publicly available and visible in the blockchain. For this purpose, however, privacy-preserving implementations (e.g., Hawk [23]) can be used, where zero-knowledge-proofs are employed in order to hide information in the blockchain.

4.5.2 Privacy and Security

The proposed protocol does not rely on a third party like [4], but uses a blockchain and smart contracts instead. Using these technologies, the four requirements introduced in Sects. 4.3.1 and 4.3.4 are fulfilled, namely:

- **Transparency.** All calculations are transparent to the smart meter and the utility providers. This is guaranteed by the properties of the blockchain [5]. Once data is written to the blockchain, it cannot be changed and therefore prevents adversaries from forging matching results.
- **Verifiability.** Calculations and the results are verifiable by all participants. This is guaranteed, since the results of smart contracts are stored in the blockchain [6, 23] and can be verified through recomputing the blocks.
- **Reliability.** The matching does not rely on a single party, but the computation is decentralized and can be performed by any peer in the blockchain [6].
- **Privacy.** The privacy-preserving properties stem from [4], with one exception that will be discussed below. Since the blockchain stores data permanently and publicly available, additional features such as a commitment scheme and private blockchains are required [23].

In the following, the privacy properties of this protocol are discussed. While most of the privacy analysis is already conducted in [4], the replacement of the third party by a public blockchain requires a discussion of privacy of previously non-public data.

This protocol is privacy-preserving as none of the participating entities has access to the original load profiles of the others, but only to the embedded load profiles. In [4], the privacy features of the embedding approach are discussed in detail.

In [4], it is assumed that all parties involved in communication through our proposed protocol are authenticated, e.g., through X.509 certificates [39]. This is not necessary because, using a blockchain, all parties are already automatically authenticated using public key cryptography [5] where each peer in the blockchain has a unique key, also referred to as address.

Similarly, in [4], for the *initialization* and *matching* phases, it is assumed that all communication links are encrypted, e.g., by symmetric encryption such as AES [40]. In contrast, such an encryption of the communication links is not used in the proposed protocol. Instead, two techniques are applied that prevent others from breaking confidentiality. A commitment scheme is used to prevent other parties from learning embedded load profiles during the commitment phase of a smart contract. However, this does not prevent the embedded load profiles from being released once the smart contract has been evaluated, which requires to open the commitments.

In the proposed protocol as shown in Fig. 4.3, the parameters $\mathbf{A}$, $\mathbf{W}$, and Δ are only known to the smart meter and the utilities as they are in [4]. However, in the proposed protocol, the embedded load profile forecast $\tilde{\mathbf{F}}$ and the embedded template load profiles $\tilde{\mathbf{L}}_{u,l}$ are publicly known since they are committed into the blockchain. As a consequence, the smart meter and all utilities know the embedding parameters and all embedded load profiles. Thus, privacy boils down to the question how much information can be inferred from both, the embedding parameters and the embedded load profiles.

Since the focus of this work is to show that blockchains can be used in principle for the protocol originally proposed in [4], the privacy analysis of this aspect is left as future work. However, it should be noted that privacy-preserving smart contracts as described in [23] can be used at the cost of reduced transparency. This way, embedded load profiles remain completely private even after the execution of a smart contract, thus solving this issue.

In Sect. 4.3.2.1 is discussed that in [4] the third party may collect statistics on the matching results by collecting the indices (u^*, l^*). Principally, the same holds for the blockchain-based approach where these indices are the result of evaluating the smart contract. For the approach with a **TP**, utilities can regularly shuffle their template load profiles' indices. The same approach can be applied in the approach presented in this paper. Additionally, all parties are only identified by their **ID** or address in the blockchain. This pseudonymity in combination with the shuffling of the indices prevents other parties from learning such statistics. Furthermore, bulk probing of either the smart meter or the utility providers is a similar issue, which is discussed in detail in [4].

In the *oblivious transfer* phase, which is identical to [4], no additional encryption is necessary. The security properties of oblivious transfer are not discussed in this analysis, but can be found in literature, e.g., [10, 24, 25], since this analysis mainly focuses on the privacy impact of the information that is transferred.

4.5.3 Comparison to Related Work

In this section, the proposed blockchain-based approach is compared to existing approaches. Figure 4.4 shows a comparison of three approaches: the presented approach, the approach relying on a third party, which we extend in this paper, and an approach using homomorphic encryption. This section extends the discussion originally presented in [4].

The core purpose of the presented approach is to find similarities, i.e., nearest neighbors, of time series by Euclidean distance computation while preserving the privacy of all actors. In [9, 17, 41], approaches based on additively homomorphic encryption, e.g., the Paillier cryptosystem [42], are discussed which also allow calculating Euclidean distances. Thus, they can be used as an alternative to nearest-neighbor embeddings as proposed in this paper and in [4]. Therefore we compare our approach to the latter. Figure 4.4 (left) shows a simplified version of our proposed scheme in this paper, (center) of the scheme proposed in [4] and (right) approaches using homomorphic encryption.

In the following, the approach using homomorphic encryption is described. Furthermore, the embedding-based approach from [4] is compared to the one using homomorphic encryption. Finally, our proposed approach, which is similar to the embedding-based approach is compared to the latter and the communication overhead is discussed.

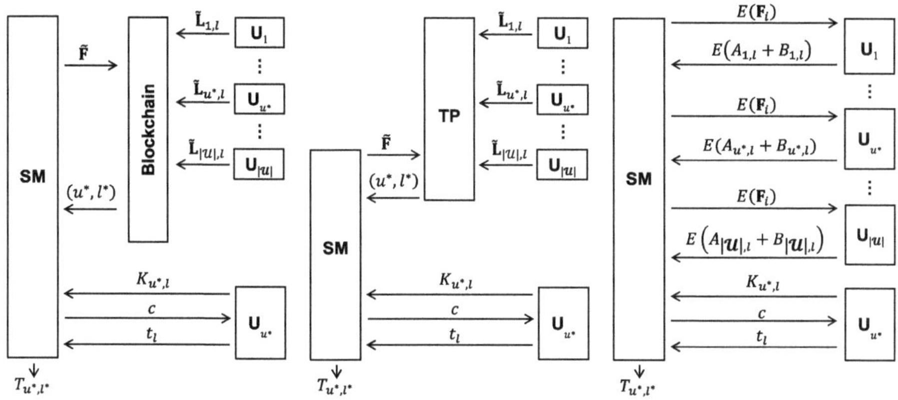

Fig. 4.4 Comparison of three distinct approaches for privacy-preserving tariff matching. From *left to right*: (1) blockchain-based approach as presented in this work; (2) embedding-based approach that relies on a third party for matching as presented in [4]; and (3) approach that uses homomorphic encryption proposed in [9, 17] and discussed in [4]

In the homomorphic encryption-based approach the smart meter communicates with each utility separately sending a load profile forecast encrypted with an additively homomorphic scheme. Each utility responds with a partial result for each template load profile as shown below. From this, the smart meter can calculate the Euclidean distance in order to retrieve the index (u^*, l^*) of the best-matching template load profile. Finally, the smart meter and the utility u^* run the oblivious transfer protocol identically to our approach in order to retrieve the tariff T_{u^*,l^*}.

Like our proposed protocol, this homomorphic protocol comes without the need of a third party. By contrast, fetching the template load profiles as well as the distance computation itself is performed by the smart meter. All encryptions are performed with the public key, decryptions can only be performed by the smart meter, which is the only party owning the private key.

The Euclidean distance between the forecast load profile and each of the template load profiles of all utilities must be computed by using an additively homomorphic cryptosystem [9] in an identical manner, i.e., with the same modulus n of Paillier's cryptosystem. Therefore, for the sake of readability, $\mathbf{L}_i$ is written instead of $\mathbf{L}_{u,l,i}$ and A, B are written instead of $A_{u,l}$ and $B_{u,l}$. The goal of the protocol is to privately compute

$$||\mathbf{F} - \mathbf{L}||_2 = \sum_i \mathbf{L}_i^2 - \sum_i 2\mathbf{F}_i\mathbf{L}_i + \sum_i \mathbf{F}_i^2 =: A + B + C. \tag{4.5}$$

First, the smart meter submits its encrypted load profile forecast values $E(\mathbf{F}_i)$ directly to each utility. As a single load profile forecast value is much smaller than the modulus, data packing is used to better exploit the input domain. Therefore, not a single value is encrypted, but all k values of the load profile forecast are packed, encrypted, and sent as one message. Data packing is achieved by shifting values to a certain bit range, such that for all operations the value remains within that range [16], e.g., for k values and a range of b bits $v = v_1|v_2|\ldots|v_k = \sum_{i=1}^{k} v_i 2^{b(i-1)}$.

Using this encrypted load profile forecast and exploiting the homomorphic properties $E(x + y) = E(x)E(y)$ and $E(x)^c = E(cx)$, the utility can compute and send back the partial result

$$E(A + B) = E\left(\sum_i \mathbf{L}_i^2\right) \prod_i E\left(\mathbf{F}_i\right)^{-2\mathbf{L}_i} \tag{4.6}$$

to the smart meter. Exploiting the homomorphic property again, term C can be added and the smart meter gets the Euclidean distance by

$$||\mathbf{F} - \mathbf{L}||_2 = D\left(E\left(\sum_i \mathbf{F}_i^2\right) E(A + B)\right) \tag{4.7}$$

When following the above protocol, neither the utility knows the smart meter's load profile forecast nor does the smart meter know the utility's template load profile. In addition, the inner product $\mathbf{F} \cdot \mathbf{L}$ is never revealed to any of them.

Now, the communication need, i.e., bandwidth requirement, is calculated and compared with this paper's solution. In the first step of the protocol, the smart meter needs to send k packed and homomorphically encrypted load values F_i to each of the $|\mathcal{U}|$ utilities. By encrypting a single value with the Paillier cryptosystem, a plaintext $p \in \mathbb{Z}_n^*$ results in a ciphertext $E(p) \in \mathbb{Z}_{n^2}^*$ leading to an expansion of the bit size by a factor of 2.

The second message is the encryption of

$$A + B = \sum_i \mathbf{L}_i^2 - \sum_i 2\mathbf{F}_i\mathbf{L}_i \in \left[-\sum_i F_i^2, \sum_i L_i^2\right] \tag{4.8}$$

where the lower limit follows from the fact that

$$||\mathbf{F} - \mathbf{L}||_2 = A + B + C \geq 0 \Rightarrow A + B \geq -C. \tag{4.9}$$

Since squaring of a number leads to an expansion of 2, both, the lower and upper limits need a maximum of $k \cdot 2s$ bits, where s is the bit size of a single load value F_i. Therefore, $A + B$ needs $2 \cdot 2ks = 4ks$ bits. Because of subsequent data expansion by a factor of 2 due to encryption, finally, the ciphertext fits into $8ks$ bits, which would require a modulus of $8ks$ bits size for the Paillier cryptosystem with modulus n. If the modulus is smaller (see below for a practical example), the message can be split into multiple messages, the number of which is $\lceil \frac{8ks}{2n} \rceil = \lceil \frac{4ks}{n} \rceil$.

As a practical example consider $k = 96$ values of a day profile arising from a 15-min measurement interval, a bit size of $s = 16$ and $|\mathscr{U}| = 5$ utilities, each having $|\mathscr{L}_u| = 20$ template load profiles. Therefore, a template load profile (as well as load profile forecast) is of size $4ks = 4 \cdot 96 \cdot 16 = 6144$ bits. According to latest NIST recommendations [43], a Paillier modulus of $n = 2048$ bits (expanding to 4096 bits) is chosen, which requires three messages of that size, since $\frac{8ks}{2n} = \frac{4ks}{n} = 3$.

For the homomorphic encryption approach, the smart meter sends its homomorphically encrypted load profile forecast to each of the $|\mathscr{U}|$ utilities. As described above, sending one load profile forecast requires three messages of 4096 bits size after encryption. Sending all load profile forecasts in this step therefore requires $3 \cdot 4096|\mathscr{U}|$ bits.

In addition, the *oblivious transfer* step at the end requires one message of 256 bits size. Thus, the total bandwidth for sending needed by a single smart meter is 7.53 KiB (rounded to two decimal places).

Each utility responds to the requesting smart meter with a partial result [see Eq. (4.6)] for each of the $|\mathscr{L}_u|$ template load profiles, as described above. From this, the smart meter calculates the Euclidean distance. One template load profile requires $3 \cdot 4096$ bits, as described above. One utility therefore sends $3 \cdot 4096|\mathscr{L}_u|$ bits. Thus, $|\mathscr{U}|$ utilities send $3 \cdot 4096|\mathscr{L}_u||\mathscr{U}|$ bits, which are received by the smart meter.

In addition, the *oblivious transfer* step requires sending $|\mathscr{L}_u|$ messages of 256 bits size. The smart meter thus receives a total of $3 \cdot 4096|\mathscr{L}_u||\mathscr{U}| + 256|\mathscr{L}_u|$ bits $= 150.63$ KiB (rounded to two decimal places). The total required bandwidth is shown in Fig. 4.5 (solid and dashed black lines).

With the embedding approach, the smart meter needs to send its load profile forecast consisting of m bits of data to the third party. All $|\mathscr{U}|$ utilities need to send $|\mathscr{L}_u|$ template load profiles of m bits each, totaling $|\mathscr{U}||\mathscr{L}_u|m$ bits. For the communication of the best-matching indices (u^*, l^*), $s = 16$ bits $= 2$ B should suffice.

Figure 4.5 shows the bandwidth requirement for the embedding approach presented in [4] (solid and dashed red lines, as well as dash-dotted blue line) for various embedding dimensions m and a third party. As discussed in [4], $m = 8192$ (dashed-dotted gray line) is a reasonable choice, resulting in an overhead of 100 KiB.

The bandwidth for sending and receiving required by smart meters is orders of magnitudes smaller for the embedding protocol than for the protocol with homomorphic encryption. This is important, since in practical scenarios the bandwidth connecting smart meters with other parties is likely to be low [44], especially when using power-line communication (PLC). While the total communication amount for the embedding method can even be higher than for the homomorphic method, most of it is needed for the communication from utilities to the third party where a much better connection is likely.

The approach based on homomorphic encryption does not require a third party to perform the distance calculation, since either of the participants is involved exactly once in exchanging messages and can limit the rate of requests in order

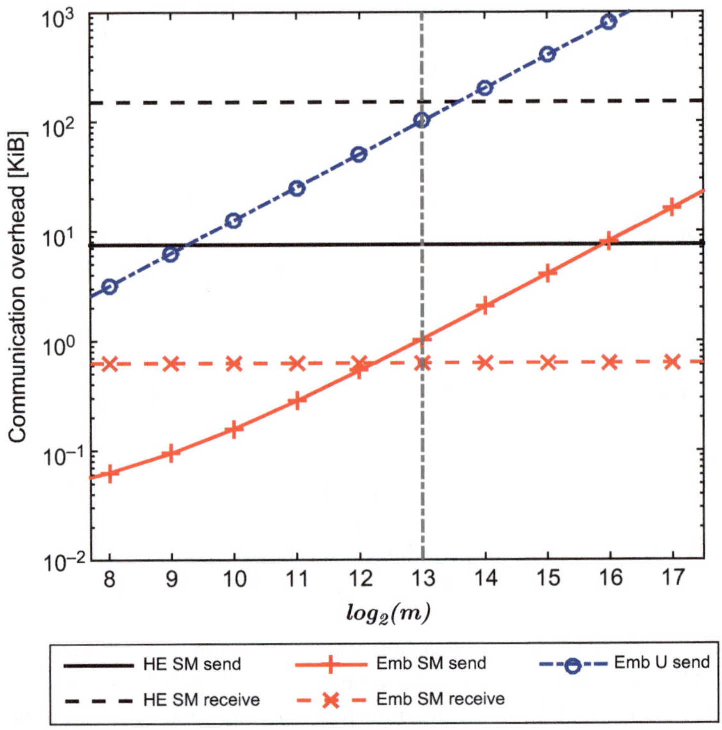

Fig. 4.5 Comparison of the required bandwidth for the protocol based on homomorphic encryption and the one from [4] with variable embedding dimension m. The bandwidth needed by the smart meter is considerably lower with the embedding protocol (denoted *Emb*) than with the protocol using homomorphic encryption (denoted *HE*). Utilities only need to send data to the **TP** using the embedding protocol (*blue, dashed-dotted*)

to prevent chosen-plaintext attacks to learn about load profiles. However, the smart meter—which is usually a device with only limited computational capabilities—has to perform the distance computation for every template load profile from every utility, which results in a total of $\sum_u |\mathscr{L}_u|$ distance computations. This is likely to be impractical for a device with low computational capabilities like a smart meter and thus another disadvantage compared to our approach.

However, the homomorphic approach calculates all distances exactly. This is not the case in the profile matching approaches presented in this paper and in [4]. In summary, the smaller overhead in data expansion comes at the cost of only near-perfect matching. However, the accuracy depends on m which can be chosen appropriately as shown in [4].

Finally, in terms of bandwidth the proposed protocol is nearly identical to the embedding-based protocol. As shown in Fig. 4.4, the third party **TP** is replaced by a blockchain, but the same messages need to be sent.

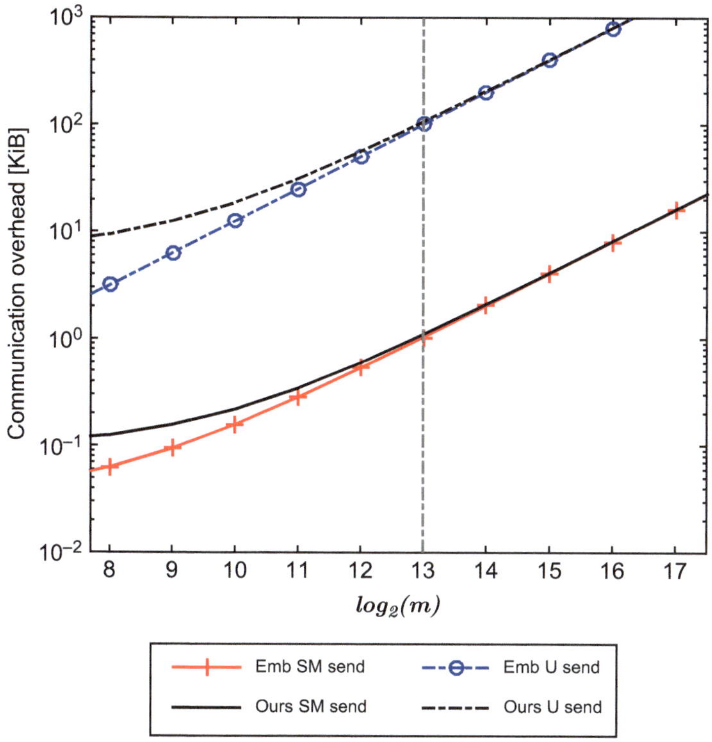

Fig. 4.6 Comparison of the required bandwidth for our proposed protocol and the one from [4], which both are based on embeddings. This figure uses the same scaling as Fig. 4.5 with variable embedding dimension m. The bandwidth needed for our proposed protocol (denoted *ours*) is slightly higher for the smart meter and utilities (due to the commitments) than with the protocol using a **TP** (denoted *Emb*) that is not required in our protocol. While the required bandwidth for sending is slightly higher, both protocols perform equally well for receiving (not depicted)

However, there is an additional overhead due to the use of commitments—depending on the size of the hash y as discussed in Sect. 4.3.3. Assuming $y = 256$ for SHA-2, each commitment requires 512 bits in total—256 bits for the hash as the commitment and 256 bits for the random number in the opening. The smart meter sends these 512 additional bits once for the load profile forecast, whereas each utility sends 512 additional bits per template load profile, resulting in a total required bandwidth of $512|\mathscr{L}_u|$. As shown in Fig. 4.6, this additional overhead becomes negligible for large values of m. For $m = 8192$, as argued practically in [4], the sending overhead is approximately 6% for both, **SM** and **U**. The receiving overhead is identical to the embedding approach.

4.6 Conclusion

We described a load profile matching protocol which enables tariff decisions in smart grids using blockchains and smart contracts. Our protocol finds the best-matching tariff for a customer with 93.5% accuracy, while ensuring transparency, verifiability, and reliability. The proposed protocol outperforms homomorphic encryption and has comparable communication overhead to previous related work [4] without relying on a third party. Instead of a single third party, a blockchain and smart contracts are used which allow for decentralized, privacy-preserving tariff-decisions in the smart grid. As in [4], the privacy of all participants depends on the usage of embedding transformations with one exception that can be mitigated by the use of privacy-preserving smart contracts [23].

Future work will focus on an actual implementation in Solidity and an evaluation of the costs associated with the evaluation of the proposed smart contract, especially when using privacy-preserving smart contracts as proposed in [23].

Acknowledgements The financial support by the Austrian Federal Ministry of Science, Research and Economy and the Austrian National Foundation for Research, Technology and Development is gratefully acknowledged. Funding by the Federal State of Salzburg is gratefully acknowledged. The authors would like to thank their partner Salzburg AG for providing real-world load data.

References

1. Karg L, Kleine-Hegermann K, Wedler M, Jahn C. E-Energy Abschlussbericht—Ergebnisse und Erkenntnisse aus der Evaluation der sechs Leuchtturmprojekte, Bundesministerium fr Wirtschaft und Technologie (German federal ministry for economy and technology), Tech. Rep., 2014, in German [Online]. http://www.digitale-technologien.de/DT/Redaktion/DE/Downloads/ab-gesamt-begleitforschung.pdf?__blob=publicationFile&v=4
2. Lisovich M, Mulligan D, Wicker S. Inferring personal information from demand-response systems. IEEE Secur Priv. 2010;8(1):11–20.
3. McKenna E, Richardson I, Thomson M. Smart meter data: balancing consumer privacy concerns with legitimate applications. Energy Policy 2012;41:807–814.
4. Unterweger A, Knirsch F, Eibl G, Engel D. Privacy-preserving load profile matching for tariff decisions in smart grids. EURASIP J Inf Secur. 2016;2016(1):21.
5. Nakamoto S. Bitcoin: a peer-to-peer electronic cash system. Bitcoin.org, pp. 1–9, 2008 [Online]. https://bitcoin.org/bitcoin.pdf
6. Wood G. Ethereum: a secure decentralised generalised transaction ledger. Ethereum, Tech. Rep., 2017 [Online]. https://ethereum.github.io/yellowpaper/paper.pdf
7. McDaniel P, McLaughlin S. Security and privacy challenges in the smart grid. IEEE Secur Priv Mag. 2009;7(3):75–77.
8. Eibl G, Engel D. Influence of data granularity on smart meter privacy. IEEE Trans. Smart Grid 2015;6(2):930–39.
9. Rane SD, Boufounos P. Privacy-preserving nearest neighbor methods: comparing signals without revealing them. IEEE Signal Process Mag. 2013;30(2):18–28.
10. Kilian J. Founding cryptography on oblivious transfer. In: ACM symposium on theory of computing. Chicago: ACM; 1988. p. 20–31.

11. Mukherjee S, Chen Z, Gangopadhyay A. A privacy-preserving technique for Euclidean distance-based mining algorithms using Fourier-related transforms. VLDB J. 2006;15(4): 293–315.
12. Ravikumar P, Cohen WW, Fienberg SE. A secure protocol for computing string distance metrics. In: International conference on data mining (ICDM). 2004. p. 40–46.
13. Wong WK, Cheung DWL, Kao B, Mamoulis N. Secure kNN computation on encrypted databases categories and subject descriptors. In: Proceedings of the 35th SIGMOD international conference on management of data. 2009. p. 139–152 [Online]. http://doi.acm.org/10.1145/1559845.1559862
14. Boufounos PT, Rane S. Efficient coding of signal distances using universal quantized embeddings. In: 2013 Data compression conference (DCC). 2013. p. 251–60.
15. Cheon JH, Kim M, Lauter K. Homomorphic computation of edit distance, vol. 8976. Berlin/Heidelberg: Springer; 2015. p. 194–212.
16. Erkin Z, Veugen T, Toft T, Lagendijk RL. Generating private recommendations efficiently using homomorphic encryption and data packing. IEEE Trans Inf Forensics Secur. 2012;7(3):1053–66.
17. Rane SD, Sun W, Vetro A. Secure distortion computation among untrusting parties using homomorphic encryption. In: 2009 16th IEEE international conference on image processing (ICIP). 2009. p. 1485–8.
18. Barni M, Bianchi T, Catalano D, Di Raimondo M, Labati RD, Failla P, et al. A privacy-compliant fingerprint recognition system based on homomorphic encryption and fingercode templates. In: IEEE 4th international conference on biometrics: theory, applications and systems (BTAS 2010). 2010. p. 1–7.
19. Sadeghi AR, Schneider T, Wehrenberg I. Efficient privacy-preserving face recognition. In: Lee D, Hong S, editors. Information, security and cryptology (ICISC 2009). Lecture notes in computer science, vol. 5984. Berlin/Heidelberg: Springer; 2010. p. 229–244.
20. Kolesnikov V, Sadeghi AR, Schneider T. Improved garbled circuit building blocks and applications to auctions and computing minima. In: Garay JA, Miyaji A, Otsuka A, editors. Cryptology and network security (CANS 2009). Lecture notes in computer science, vol. 5888. Berlin/Heidelberg: Springer; 2009. p. 1–20.
21. Ben-Sasson E, Chiesa A, Garman C, Green M, Miers I, Tromer E, et al. Zerocash: decentralized anonymous payments from bitcoin. In: Proceedings—IEEE Symposium on Security and Privacy. IEEE; 2014. p. 459–474.
22. Zyskind G, Nathan O, Pentland AS. Decentralizing privacy: using blockchain to protect personal data. In: Proceedings—2015 IEEE security and privacy workshops (SPW 2015). 2015. p. 180–184.
23. Kosba A, Miller A, Shi E, Wen Z, Papamanthou C. Hawk: the blockchain model of cryptography and privacy-preserving smart contracts. In: 2016 IEEE symposium on security and privacy (SP). IEEE; 2016. p. 839–58.
24. Yao ACC. How to generate and exchange secrets. In: 27th annual symposium on foundations of computer science. Washington: IEEE Computer Society; 1986. p. 162–7.
25. Catalano D, Cramer R, DiCrescenzo G, Darmgard I, Pointcheval D, Takagi T. Provable security for public key schemes. Basel: Birkhäuser Verlag; 2005.
26. Palensky P, Dietrich D. Demand side management: demand response, intelligent energy systems, and smart loads. IEEE Trans Ind Inf 2011;7(3):381–8.
27. Caron S, Kesidis G. Incentive-based energy consumption scheduling algorithms for the smart grid. In: 2010 First IEEE international conference on smart grid communications (SmartGridComm). 2010. p. 391–6.
28. Shao S, Zhang T, Pipattanasomporn M, Rahman S. Impact of TOU rates on distribution load shapes in a smart grid with PHEV penetration. In: 2010 IEEE PES transmission and distribution conference and exposition: smart solutions for a changing world. 2010. p. 1–6.
29. Ramchurn S, Vytelingum P, Rogers A, Jennings N. Agent-based control for decentralised demand side management in the smart grid. In: The 10th international conference on autonomous agents and multiagent systems, AAMAS '11, vol. 1. Taipei: International Foundation for Autonomous Agents and Multiagent Systems; 2011. p. 5–12 [Online]. http://eprints.soton.ac.uk/271985/

30. Mohsenian-Rad AH, Wong VWS, Jatskevich J, Schober R, Leon-Garcia A. Autonomous demand-side management based on game-theoretic energy consumption scheduling for the future smart grid. IEEE Trans. Smart Grid 2010;1(3):320–31.
31. Knirsch F. Privacy enhancing technologies in the smart grid user domain. it—Inf Technol. (Thematic Issue: Recent Trends in Energy Informatics Research) 2017:59(1):13–22.
32. Gennaro R, Katz J, Krawczyk H, Rabin T. Secure network coding over the integers. In: Pointcheval D, Nguyen PQ editors. Public key cryptography (PKC 2010). Lecture notes in computer science, vol. 6056. Berlin/Heidelberg: Springer; 2010. p. 142–60.
33. Fiore D, Gennaro R, Pastro V. Efficiently verifiable computation on encrypted data. In: Proceedings of the 2014 ACM SIGSAC conference on computer and communications security, CCS '14. Scottsdale: ACM; 2014. p. 844–55.
34. Parakh A. Oblivious transfer using elliptic curves. In: 15th international conference on computing. IEEE; 2006, p. 323–8.
35. Barker A. NIST special publication 800-57: recommendation for key management—part 1: general (revised). 2016 [Online]. http://csrc.nist.gov/publications/nistpubs/800-57/sp800-57-Part1-revised2_Mar08-2007.pdf
36. Peters, GW, Panayi E. Understanding modern banking ledgers through blockchain technologies: future of transaction processing and smart contracts on the internet of money. In: Paolo T, Aste T, Pelizzon L, Perony N, editors. Banking beyond banks and money: a guide to banking services in the twenty-first century. Cham: Springer International Publishing; 2016. p. 239–78.
37. Delmolino K, Arnett M, Kosba AE, Miller A, Shi E. Step by step towards creating a safe smart contract: lessons and insights from a cryptocurrency lab. In: Financial cryptography and data security. Barbados: International Financial Cryptography Association; 2016. p. 79–94.
38. Pedersen TP. Non-interactive and information-theoretic secure verifiable secret sharing. In: Advances in cryptology (Crypto '91), vol. 91. 1992. p. 129–40
39. ITU-T, Recommendation ITU-T X.509—information technology—open systems interconnection—the directory: public-key and attribute certificate frameworks. 2012.
40. National Institute of Standards and Technology (NIST). Specification for the advanced encryption standard (AES). 2001.
41. Lagendijk R, Erkin Z, Barni M. Encrypted signal processing for privacy protection. IEEE Signal Process Mag. 2013;30:82–105.
42. Paillier P. Public-key cryptosystems based on composite degree residuosity classes. In: Stern J, editor. Advances in cryptology—EUROCRYPT '99: international conference on the theory and application of cryptographic techniques Prague, Czech Republic, May 2–6, 1999 proceedings, Lecture notes in computer science, vol. 1592. Berlin/Heidelberg: Springer; 1999. p. 223–238.
43. Barker E, Barker W, Burr W, Polk W, Smid M, Division, CS. NIST 800-57: computer security. 2012. p. 1–147.
44. Unterweger A, Engel D. Resumable load data compression in smart grids. IEEE Trans Smart Grid 2015;6(2): 919–929 [Online]. http://dx.doi.org/10.1109/TSG.2014.2364686

Chapter 5
Energy Cloud: Services for Smart Buildings

Nader Mohamed, Jameela Al-Jaroodi, and Sanja Lazarova-Molnar

5.1 Introduction

Based on a United Nations study, the urban population around the world has increased from only 746 million in 1950 to 3.9 billion in 2014, while it is estimated that this number will increase further by 66% to more than 6 billion by 2050. Moreover, there were only ten "mega-cities" with 10 million or more inhabitants in 1990, it is predicted that there will be 41 mega-cities by 2030. In addition, many other cities are rapidly increasing in population while they have very limited vacant lands. As a result, there will be more plans for vertical expansion, thus resulting in constructing many large residential and commercial buildings to support this population growth. However, buildings are responsible for a significant portion of the total energy consumption and greenhouse gases (GHG) production. For example, buildings consume around 49% of the total energy and contribute around 47% of the greenhouse gas emissions in the United States [1]. In addition, buildings in Europe consume 40% of the total energy used and contribute around 36% of the carbon dioxide emissions [2]. Due to these high numbers, both the US Department of Energy and the European Commission aim to gradually reduce the use of primary energy in buildings.

N. Mohamed (✉)
Middleware Technologies Lab., Isa Town, Bahrain
e-mail: nader@middleware-tech.net

J. Al-Jaroodi
Department of Engineering, Robert Morris University, Moon, PA 15108, USA
e-mail: aljaroodi@rmu.edu

S. Lazarova-Molnar
Center for Energy Informatics, University of Southern Denmark, Odense, Denmark
e-mail: slmo@mmmi.sdu.dk

W. Rivera (ed.), *Sustainable Cloud and Energy Services*,
DOI 10.1007/978-3-319-62238-5_5

One of the main directions of this energy consumption reduction is to utilize Information and Communication Technology (ICT) to turn regular buildings into smart buildings. This involves deploying and interconnecting various sensors, actuators, subsystems, and applying advanced and smart automation monitoring and control mechanisms to reduce energy needs in buildings. This model creates what is known as the building energy management system (BEMS) [3]. By utilizing intelligent control and computational models to automatically manage energy consumption in smart buildings, BEMS offer great opportunities to enhance this efficiency of these buildings. Due to the high benefits of such approach, huge research and industrial efforts were conducted to propose more enhanced monitoring and control mechanisms, analysis algorithms, better monitoring technologies, and more efficient devices and subsystems that can further help in reducing energy consumption in smart buildings. In addition, some governments around the world have started to enforce some regulations and policies in designing and building more energy-efficient building.

The efforts and advanced research and industrial applications for BEMS have offered some improvements. However, there are some obstacles in effectively utilizing these efforts in buildings to effectively reduce energy consumptions. Some of these obstacles are the high heterogeneity of buildings in terms of structures, usage, and equipment installed; the high complexity of identifying the best monitoring and controls mechanisms for buildings in various situations; the inflexibility in upgrading to new control mechanisms that can provide better results; the lack of analysis tools that help in indenting any new energy issues; and the high costs associated with analyzing installing, and upgrading such systems. Some of these obstacles may be leveraged by combining the efforts across multiple buildings and addressing the common issues involved. However, this requires more technical resources and higher computing power to achieve effectively.

In the meantime, cloud computing has been introduced to solve many issues in industry in flexible and cost-effective manners. It is an on-demand computing model that provides highly scalable computation power and data storage capacities [4]. Furthermore, it can provide services to solve various issues in different domains including government, finance, manufacturing, transportations, healthcare, and online entertainment [5]. They can also solve some issues in the energy domain including energy efficiency in smart buildings which is our focus in this chapter. We will discuss and review utilizing cloud computing for providing energy-related services for enhancing energy efficiency in smart buildings.

This chapter is organized as follows. Section 5.2 discusses background information about smart buildings while Sect. 5.3 lists and discusses current limitations in smart buildings. Section 5.4 covers some background information about cloud computing. Section 5.5 discusses integrating smart buildings with the cloud and the different issues imposed by that integration. Different possible energy cloud services that can be provided for smart buildings and their benefits are discussed in Sect. 5.6. Section 5.7 discusses the issues associated with the energy cloud. Some of the current research efforts and industrial solutions are discussed in Sect. 5.8. Finally, Sect. 5.9 concludes the chapter.

5.2 Smart Buildings

This section provides background information about smart buildings and the technologies used. There are two main approaches to save energy in buildings. The first is to use special insulation materials in the buildings' walls to help reduce heat exchange and maintain certain temperatures in the building. The second approach is to use automation techniques that implement smart algorithms to control buildings' equipment such as HVAC (heating, ventilating, and air-conditioning) systems, home and office appliances, and lighting systems. This approach requires the deployment of a building energy management system (BEMS) [3] that implements such algorithms and relies on various types of sensor nodes to monitor the current energy usage and environmental conditions and actuators to control the equipment in the building.

Smart buildings are usually equipped with a variety of sensor devices such as temperature sensors, carbon dioxide/monoxide sensors, humidity sensors, duct sensors, and occupancy sensors. In addition, they can be equipped with smart electricity meters which are electronic devices that can be connected with a computerized system through a two-way communication. Unlike regular power meters which mainly record power consumption, they can provide features such as automated meter reading (AMR) and billing, better energy utilization, detection of energy losses, early warning of blackouts, and rapid detection for energy supply disturbance [6]. The installed sensors and smart meters report their observations and measurements to the BEMS, which based on predefined objectives like required energy savings and comfort levels of the occupants makes decisions to control and adjust the operations of the equipment. Based on these decisions, the BEMS sends singles to the actuators to adjust the settings of the equipment in the building.

BEMS is considered a cyber-physical system (CPS) that integrates both the cyber world with the physical world and provides useful interactions between both worlds. In smart buildings, the physical world components are the buildings' spaces, HVAC systems and other equipment, energy supply, atmosphere conditions such as current temperature, lighting, and ventilation, and building residents. The cyber world components are the sensing devices that supply the data about the physical world components; the software that implements the monitoring and control algorithms; and the hardware components that execute the software decisions. The main three steps of BEMS, as in any other CPS application, are monitoring the physical environment through sensors; making decisions for adjustments and enhancements based on the collected information, historical information, and predefined objectives; and applying these decisions through actions executed by some actuators. These three main steps are connected in a closed loop as shown in Fig. 5.1.

Some examples of control mechanisms used by BEMS to save energy are automatically switching off area lights when workers leave that area, change the heating or air-conditioning levels based on current occupancy levels and external temperatures, and send employees' computers to standby or energy saving mode when not in use [7]. As BMES rely on computing capabilities that allow for

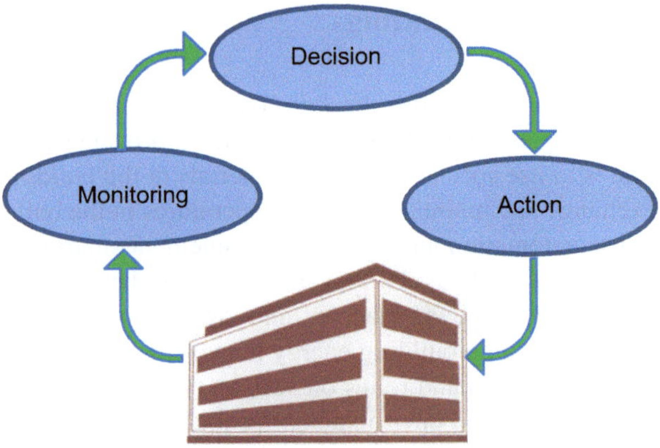

Fig. 5.1 BEMS as a cyber-physical system

developing and utilizing more sophisticated control and monitoring algorithms to optimize the controls within a building. These sophisticated algorithms may be based on genetic algorithms [8], neural networks [9], empirical models [10], weighted linguistic fuzzy rules [11], and simulation optimization [12]. In addition, more sophisticated integrated control systems that build a knowledge base and utilize it in intelligent systems that use information on occupancy predictions, optimized fuzzy controllers, and genetic algorithms [13] are proposed, developed, and successfully evaluated [14]. Furthermore, more approaches that provide better and more cost-effective decision making for BEMS are proposed. Examples of these approaches are using rule sets [15], multi-agent control [16, 17], supervisory control and data acquisition [18], scheduling-based real-time energy flow control strategy [19], and energy prices-based control [20].

5.3 Limitations with Current Smart Buildings

Although BEMS can enhance energy efficiency and reduce energy costs in smart buildings, there are some limitations with such systems. In the following, we list some of them:

- Updating BEMS control models: There are enormous and continuous advances in new and improved control mechanisms and algorithms for efficient energy consumption in smart buildings. However, it is very difficult to update the current BEMS with these mechanisms and algorithms. Some control mechanisms and algorithms could offer improved results, yet they need to be properly installed, tested, and configured to accomplish their objectives, which is generally a long, difficult, and costly process. These steps require analysis, planning, and long intervals of time to deploy and evaluate, which may not be possible as usually each smart building has a single BEMS which is continually in use and cannot be easily taken offline. In addition, this usually requires technical specialists to handle the upgrades including installation, tuning, and evaluation.

- Including new sensors and other devices: Most current BEMS are built to deal with specific types of sensors and devices. There is continuous improvement in sensor technologies and energy-efficient devices. Connecting and utilizing these new devices may require adding software models that can effectively operate them as well as altering the system's configurations. This makes the procedure of effectively utilizing new sensors and devices very challenging.
- Fault detection and diagnosis: One of the issues facing smart buildings are faults in the energy subsystems. As energy subsystems in smart buildings rely on different software and hardware components such as sensors, actuators, networks, and other devices, faults in these components may create faults in the subsystems. One example is when a temperature sensor fails and allows air conditioners to continuously work. Another example is a fault in a light sensor that could keep some lighting devices on without a valid reason. In both examples, more energy will be consumed without a valid reason. In some cases, going undetected for some time, faults could eventually accumulate and lead to a significant increase in energy consumption. Furthermore, some of these faults are hard to detect so the increases in consumption could run unnoticed for extended periods of time. It is estimated that faults contribute between 15% and 30% of the energy consumption costs [21]. As a result, Fault Detection and Diagnostics (FDD) is very important in smart buildings. It has been shown that by deploying automated FDD as part of the BEMS, the operational costs of buildings can be significantly decreased [21]. However, this requires advanced diagnostic approaches and automated algorithms that are not available in current BEMS.
- Limited storage and processing capabilities: Some advanced monitoring and control mechanisms require a large storage capacity for storing building energy information collected over extended periods of time. This information is processed using complex and compute intensive algorithms that optimize energy consumption in smart buildings. However, most of the current BEMSs have limited storage and processing capabilities, thus limiting their capabilities and effectiveness.
- Separated smart buildings: Multiple separated BEMSs units are usually used to control energy efficiency in multiple buildings. These buildings can be very similar in their location, design, usage, and operating conditions. However, current BEMS deployments are independent for each building and do not integrate multiple buildings. With separated BEMS for each building, it is very difficult to collect common observations and good experiences that can be utilized for better energy efficiency among these multiple buildings. In addition, it is difficult to build a unified knowledge base to collect and organize energy consumption data from these similar buildings that can be used for future improvements and planning.
- Integration with other systems: BEMSs operations and capabilities can be enhanced by integrating them with other systems such as the smart grid [22] or with renewable energy systems, such as solar energy systems, available for the building. This could enable reducing energy costs in smart buildings by utilizing advanced techniques and control algorithms such as applying energy

prices-based controls [20]. However, this integration is not always smoothly achievable as it requires upgrades, communication connections, and amendments in the installed BEMS.

5.4 Cloud Computing

This section provides background information about cloud computing. Cloud computing is an on-demand computing model that eliminates or reduces the need for companies and organizations to own in-house and high-cost software, hardware, and network infrastructures. They also eliminate the need for having highly skilled technical professionals to support them [5]. Instead, companies and organizations can rely on other companies, referred to as cloud companies (or cloud service providers), to offer them replacement solutions as a set of services to be used for their Information Technology (IT) needs. Cloud companies offer various services to interested consumers at reasonable and lower costs [4]. National Institute of Standards and Technology (NIST) described Cloud Computing [23] as a paradigm allowing users to access shared configurable computing resources. The NIST definition is based on five main characteristics:

- Users can automatically benefit from the wide-range of Cloud services without communicating with the service providers.
- Standard interfaces and protocols are used to access the computing services and resources over the Internet.
- Cloud services follow a multi-tenant model allowing resources to be pooled and shared among users.
- Computing capabilities can be quickly scaled in or out based on the users' varying demands.
- Users pay for utilized computing capabilities based on a pay-per-use model.

The cloud computing paradigm follows three main service models which can be offered to the users: Software-as-a-Service (SaaS), where a user can utilize application software (services) provided by the Cloud service provider and run on the Cloud infrastructure; Platform-as-a-Service (PaaS), where a user can develop and deploy applications onto Cloud platform infrastructures; and Infrastructure-as-a-Service (IaaS) where a user can benefit from the computing and storage resources offered and managed by a Cloud service provider. Consumers can utilize one or more of these models to satisfy their IT needs.

With the advances of technologies in the Internet of Things (IoT) [24], it is possible to connect different physical devices and allow them to exchange data over the Internet. Examples of these physical devices are cars, machines, buildings, sensors, and actuators. In addition, IoT can be integrated with Cloud Computing into what is called the Cloud of Things (CoT) [25]. This integration offers an unprecedented opportunity to create a wide array of applications including applications to reduce energy consumption in smart buildings.

5.5 Energy Cloud for Smart Buildings

There are two main approaches to build an energy cloud for smart buildings: cloud-based BEMS and cloud-enabled BEMS. In this section we describe both approaches.

5.5.1 Cloud-Based BEMS

In this approach, all buildings' subsystems, sensor nodes, and actuators that control buildings equipment are directly connected to the energy cloud through the Internet. The energy cloud provides the functions of BEMS for smart buildings. In this case the energy cloud will replace the buildings' BEMSs and provide real-time controls for energy efficiency to all connected smart buildings. Although, this approach can provide high flexibility for designing and offering the energy cloud services, there are a number of disadvantages:

- It requires having reliable and high quality communication capabilities between the participating smart buildings and the energy cloud to support real-time controls. However, this cannot be ensured with the Internet as it only provides best-effort communication services.
- It is very difficult to implement cloud services that deal with the details of large variety of sensors and other devices.
- It will generate high communication traffic on the network and will be difficult to scale for a large number of smart buildings.
- There are high security risks associated with this approach as both the energy cloud and the smart buildings can be vulnerable for intrusion and security attacks.

5.5.2 Cloud-Enabled BEMS

The second approach is to use cloud-enabled BEMSs [26] and connect smart buildings with the energy cloud through these cloud-enabled BEMSs as shown in Fig. 5.2. In this approach, the cloud-enabled BEMSs will function as regular BEMSs in smart buildings. All real-time controls of the equipment in the building will still be done locally as they are done by regular in-house BEMSs. The local BEMS connect to the energy cloud for special services. The energy cloud services will mainly collect information from smart buildings, find optimization updates to tune the energy control mechanisms in these buildings, and provide software and configuration updates for local cloud-enabled BEMS periodically. These updates will enhance both energy efficiency and occupants' comfort levels in the participating smart buildings.

Using cloud-enabled BEMSs can provide a number of advantages:

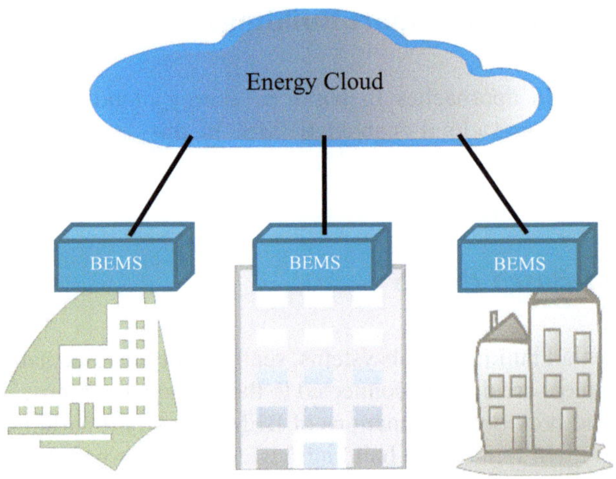

Fig. 5.2 Integrating smart buildings with the energy cloud

- Providing low-latency control services as all real-time monitoring and control mechanisms will be done locally within the building's BEMS.
- Providing better scalability support as there is no need to use the energy cloud for detailed and continuous monitoring and control.
- Providing more efficient communication between smart buildings and the energy cloud as information exchange can be managed better through periodic updates.
- Providing better reliability controls for the smart buildings as cloud-enabled BEMSs can still function even if there is no communication with the energy cloud for some periods.
- Providing better security protection as both the energy cloud and cloud-enabled BEMSs can implement some security mechanisms to protect both sides.

5.6 Energy Cloud Services

The energy cloud with its scalable data storage and processing power can provide a number of energy-related services for smart buildings. The energy cloud functions as illustrated in Fig. 5.3. It collects different types of information from the connected smart buildings and from other systems such as smart grids and weather systems. This information is validated, filtered, and stored. This information is then used to update a knowledge-base system that builds knowledge experiences for energy consumption in smart buildings. This knowledge can be utilized by energy services on the cloud to enhance the operations of the building energy management systems for better energy efficiency in smart buildings. These enhancements include improving the process of monitoring, decision making, and actions for better energy efficiency in smart buildings. The energy cloud can provide a number of services for smart buildings including:

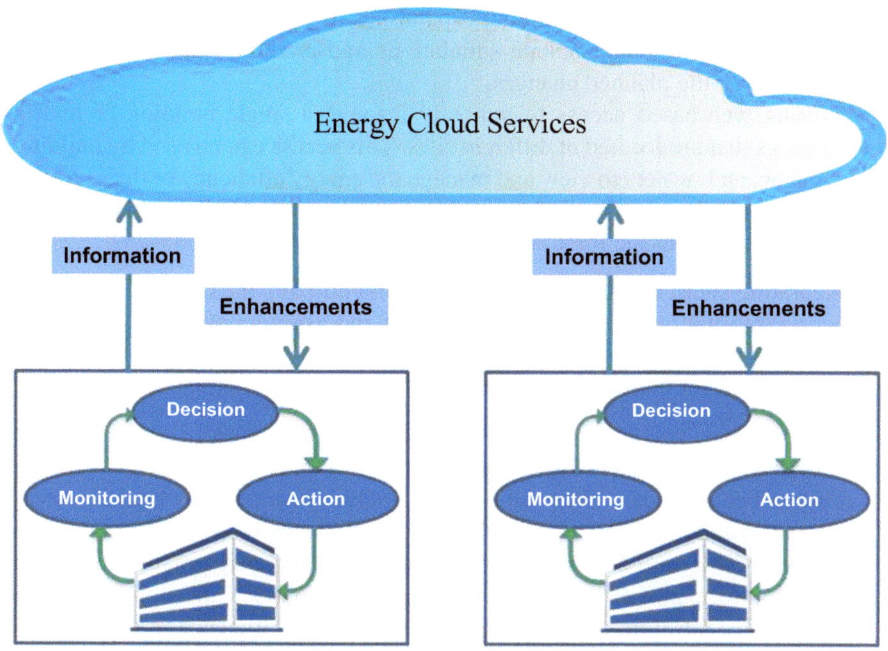

Fig. 5.3 Energy cloud services collect information from multiple smart buildings and enhance building energy management systems

- Providing the required BEMS software models that implement various monitoring and energy control algorithms.
- Providing appropriate configuration services that are needed by the monitoring and control software models executing on the BEMSs for more efficient energy monitoring and control.
- Providing regular updates for the BEMSs software models that can deal with new energy subsystems, devices, sensor nodes, and actuators; or that contain more energy-efficient control models using better or more advanced algorithms.
- Configuring the connected BEMSs to collect a certain type of information about energy usage and environmental conditions in the participating smart buildings. This information can be used to enhance the decision-making process in the energy cloud, which will help improve the control software models.
- Building a knowledge-based system for energy issues and usage information for multiple smart buildings. This system can be used as a main source for planning, optimizations, and future enhancements.
- Offering advanced data analytics services that use the knowledge-based system to find new issues, problems, and observations in the energy systems; and to provide forecast services for different energy issues in smart buildings.
- Offering optimization and evaluation services for selecting the best control mechanism for different smart buildings. The optimization services may use

machine learning and data mining capabilities to enhance the results, while the evaluation services can include simulations and predictive analysis to provide estimates for the planned changes.

- Offering web-based access to remotely manage a single building or multiple buildings that are located at different sites. This access can be used by buildings' managers and owners to view and manage the energy efficiency of their buildings and to assess the outcomes of the various services available using benchmarking services, energy usage forecasts, and unusual patterns detection.

Integrating smart buildings with the cloud and utilizing the cloud helps facilitate collecting information from multiple buildings and using it collectively to improve operations in each of these buildings. It provides common features and functions for multiple BEMS for multiple smart buildings and will result in many benefits:

- *Better energy efficiency*: The cloud can collect data from smart buildings including energy meters data, equipment status data, environmental data, and resident behavior data. This data can be analyzed by certain cloud services to provide evaluations for equipment energy efficiency, monitoring and control processes efficiency, renewable energy utilization efficiency, and overall building energy performance efficiency. Based on these evaluations, the cloud can alter or change the control modules to achieve better energy efficiency in the connected buildings.
- *Better fault detection and diagnosis*: The Cloud can offer services for timely fault detection and diagnosis for smart buildings. It can provide efficient processes for collecting, storing, and analyzing smart building data to detect and report faults in smart buildings' energy systems. There are three main categories of methods for FDD problems: model-based, data-based, and hybrid approaches that combine both data- and model-based approaches. The model-based approaches are feasible for small buildings or simple systems as they are difficult to design and use for large and complex systems such as large smart buildings. Data-based models are easier to develop, but their accuracy to a large extent depends on the availability and quality of data. The scalable storage and processing capacity of the Cloud can effectively enable data-based FDD approaches. Faults can be found by applying data mining and analysis methods to find discrepancies in the data or unusual patterns. This relies on building the knowledge-based system that collects the data from the different participating smart buildings. The knowledge is accumulated from data and experiences collected from multiple smart buildings and in some cases from multiple similar buildings that have similar structures, equipment, and usage models. This helps in detecting unknown common faults faster among these similar buildings [27]. As a result, the associated increase in consumed energy due to these faults can be avoided earlier.
- *Flexibility*: Integrating smart buildings with the cloud will provide endless flexibilities and capabilities to enhance energy consumption in smart buildings. The cloud will provide huge storage and processing capabilities, which will be utilized to collect all the data necessary to support advanced analysis and optimization services. It will also enhance the capability of selecting and applying

the monitoring and control of energy consumption methods that can best suit each individual smart building. One example is that the cloud can add virtual sensors for more accurate monitoring. Virtual sensing is an approach to provide logical sensors that provide economical alternatives to costly physical sensors. These can provide inexpensive extra logical sensing capabilities while expanding the data collection process for better smart buildings energy management. A virtual sensor utilizes information available from other lower cost or more widely available physical sensors such as fixed physical sensors and other mobile sensor devices like smartphones to calculate an estimate of the quantity of interest [28] and upgrade any modules in the energy systems in connected building for better results. The cloud can easily facilitate implementing and deploying new enhanced energy monitoring, analysis, and control methods that will be developed in the future for smart buildings.

- *Reduced cost*: Although, the initial cost of integrating smart building with the cloud can be high, the long term energy savings will exceed the initial cost. With advanced cloud services that can help analyze any energy issues in smart buildings and propose alternative solutions in a timely manner will result in significant savings resulting in a reduced net cost. The cost can also be significantly reduced with large buildings that have many duplicated units and functions or if shared among multiple buildings with similar operations and characteristics.
- *Enhanced Occupants Comfort*: Buildings are being designed and built for occupants. Therefore, occupants comfort is the most important performance measure for buildings. The cloud can enhance occupants comfort by learning about various control strategies and utilizing them for optimizing the living conditions for the occupants. Occupants continuously provide feedback, and the more feedback there is, from buildings in versatile locations with various weather conditions, the better suited can be the control in the BEMS. The globalization, which also implies that occupants of a building can be of totally different backgrounds, can also be supported by taking this into consideration for better occupants comfort, as there will be more data available.
- *Better management*: As a cloud-based operation, building management tools can be made available remotely. Owners of one or multiple buildings can easily and collectively manage these buildings using a web-based system. Access to management tools will allow owners and managers to view, compare, and update information without having to be in a specific location.
- *Support for research*: The stored energy information from multiple smart buildings can be utilized to support advanced research for further enhancing energy efficiency in these or other buildings. The collected data can also be used to find new patterns or trends in energy consumption in smart buildings and it can also be used to test and validate newly designed approaches.
- *Support for strategic planning*: Decision makers can use the energy cloud data for planning and budgeting for energy services. It can also be integrated with other Cloud services to help in forecasting future energy needs for a specific building or a set of new buildings. This will help, for example, in planning the

needs for energy supply equipment and the best utilizing of renewable energy for the buildings. Long term planning can be done based on more accurate models that will allow for informed decisions on energy needs and alternatives available.

- *Support for integration with other systems*: Most advanced systems are usually designed to integrate with the cloud. Integrating smart buildings with the cloud will enable their integration with other systems when needed. The cloud can play a broker role among smart buildings and other systems. An example of this integration is between the smart grid and smart buildings. Smart buildings can, for example, use information about high and low energy load periods from the smart grid and use it to enhance appliances' utilization times and to schedule non-essential energy consuming functions to low energy load periods. The information flow between the smart grid and smart buildings can be efficiently handled by the cloud. This allows for maximizing energy efficiency through the energy market connection [22]. In addition, the cloud can provide management services for smart buildings to work with other energy systems on the cloud to find best dynamic management mechanisms based on available optimal price options and energy alternatives for reducing the total energy costs.
- *More accurate building models*: Smart buildings are very complex systems, whose models are very complicated to derive. The cloud can support the development of more accurate models as data from many buildings can be analyzed simultaneously and, through this, many complex relationships can be derived. More accurate models can further support and enhance a number of other optimization processes and support the design process for enhancing existing buildings or designing new ones.

Table 5.1 provides a summary of the current BEMS limitations discussed in Sect. 5.3 and how the energy cloud can solve these limitations based on the list of services discussed in this section. The table also highlights the benefits of the listed energy cloud solutions.

5.7 Energy Cloud Issues

Despite the positive impact of integrating smart buildings BEMS with an energy cloud, there are many issues that need to be addressed to make this a reality. Like any large-scale distributed cloud-based system, this will require achieving specific requirements that will help design, implement, and operate such systems. Examples of these issues include:

- *Reliability*: Using the cloud for energy management in smart buildings requires assurances of high reliability for the offered services. All energy cloud services must be designed to be continuously available and error-free. All important decisions regarding energy efficiency in smart buildings should be made while maintaining the residents' required level of quality of life. Therefore, it is important to have accurate data and controls. The reliability level requirement can also

Table 5.1 Summary of current BEMS limitations and cloud solutions and benefits

Current limitation	Cloud solution	Cloud solution benefit
Difficulty updating BEMS control models	Automatic updates for control models can be provided by the cloud	Faster and more cost-effective updates
Difficulty including new sensors and other devices	The cloud can provide software drivers and configurations for new sensors and other devices needed	Faster and better utilization for new sensor and device technologies to enhance energy efficiency
Limitation of fault detection and diagnosis	Advanced and up-to-date fault detection and diagnosis techniques can be provided by the cloud	Fast and cost-effective fault detection and diagnosis processes
Limited storage and processing capabilities	Scalable and on-domain storage and processing capability	More information can be collected, stored, and processed and better knowledge can be built about smart buildings for better energy efficiency
Separated smart buildings	The cloud will simultaneously collect and process information from multiple similar and dissimilar buildings	Energy knowledge can be built faster and observations can also be found faster leading to better interpretation of data, identifying trends, and making better decisions
Limitation in integration capabilities with other systems	The integration with other systems can be achieved through the cloud	Smooth and cost-effective integration with other systems can be achieved and better utilization of available features through integration

vary depending on the type of building. For example, hospitals and emergency centers have more critical operating conditions than office buildings [29].

- *Data Management*: An energy cloud will be responsible for handling a large number of smart buildings especially in the case of a smart city. In this case, big data generated from collecting all the details of energy use in these buildings can be collected and stored in the cloud. This requires having specialized data management mechanisms that allow efficient storage and processing of this big data such that the system requirements of the different energy cloud services can be efficiently met.
- *Heterogeneity of Buildings*: Buildings are generally very heterogeneous in terms of structure, size, usage, conditions, applications, and equipment. Finding an optimal solution for energy efficiency for each individual building can be very challenging. In addition, creating a detailed set of services that can deal with this level of heterogeneity within a single building or across multiple buildings require a lot of effort and complicate any type of service to be offered.
- *Interoperability*: An energy cloud should be able to deal with different types of BEMS, sensors, actuators, and other energy subsystems. These can have varying

formats for exchanging messages. The main issue here is interoperability which involves defining an agreed-upon framework or some open protocols/APIs that enable easy integration between the energy cloud and smart buildings and their equipment and subsystems.

- *Security and Privacy*: As smart buildings will be connected to the cloud through the Internet, the systems of the connected smart buildings can be at risk as they are vulnerable to external attacks and unauthorized access. In addition, smart building information may include several forms of private and sensitive information such as occupancy intervals and other occupants' information. This information must be properly protected to avoid loss or misuse of data during transmission, storage, and processing. Fortunately, there are many security and privacy approaches developed recently for the cloud to prevent many types of attacks [30]. These approaches need to be investigated and customized to fit with the specific requirements of the energy cloud security and privacy challenges.
- *Safety*: Safety is the ability of a system to function despite malfunctions, and it often goes hand in hand with reliability and security. The fact that part of the data or BEMS would be featured on the cloud poses new challenges that need to be addressed. For safety to be at an adequate level, an extensive and cloud-specific risk management would be necessary.
- *Maintainability*: Unlike having energy management systems for an individual building, an energy cloud can be used for hundreds of buildings. This increases the complexity of system maintainability compared to an individual building's BEMS. The increase is mainly due to the need to consider the necessities and features of the multiple participating smart buildings. These needs can be completely different from some buildings to others. Therefore, maintenance in the cloud infrastructure, software, or platforms must be done without while considering these differences and achieved without having negative effects on any services provided to any participating smart building. In this regard and to simplify the maintenance processes, all cloud resources and provided services must be designed for easy maintainability and access to individualized updates when necessary.

5.8 Research and Industrial Efforts

Researchers and practitioners alike have invested in research and industrial efforts to solve various issues and propose different solutions to enable the energy cloud for smart buildings. In this section we discuss the recent research efforts in Sect. 5.8.1 and the industrial efforts in Sect. 5.8.2.

5.8.1 Research Efforts

There are several research efforts that introduced new contributions to the development and effective utilization of the energy cloud for smart buildings. For example, in research, Curry et al. [31] addressed the issue of linking collected buildings data in the cloud and the concern of data interoperability in such links. An architecture for cloud computing-based BEMS was proposed by Hong et al. in [32] while a cloud-enabled BEMS was proposed by Mohamed et al. in [26]. Javed et al. [33] proposed a design and implementation of cloud-enabled random neural network-based decentralized smart controller for HVAC systems. Bruton et al. developed, implemented, and tested a cloud-based automated fault detection and diagnosis tool for air handling units [34]. The authors reported that they tested the developed tool across four commercial and industrial sites and this resulted in over €104,000 annual energy savings. Lazarova-Molnar and Mohamed studied the importance of collaborative data analytics for smart buildings in [27] and proposed a framework for collaborative cloud-based fault detection and diagnosis in smart buildings in [35]. They also proposed a smart building diagnostics as a service on the cloud in [36]. Al Faruque and Vatanparvar [37] proposed an energy management-as-a-service for home energy management and microgrid-level energy management. Sequeira et al. [38] investigated the benefit of utilizing the energy cloud to monitor and analyze energy consumption in multiple industrial sites.

Some web-based solutions were proposed to offer better user interfaces, access, and controls for building energy management. Motegi et al. [39] proposed web-based energy information systems for energy management and demand response in commercial buildings. In addition, Papantoniou et al. [40] proposed a web-based energy management and control system to optimize the mechanisms that are implemented in existing BEMS.

5.8.2 Industrial Efforts

Countries like Denmark and the UK have started to link the smart meters in buildings and facilities in some areas to cloud-based solutions to offer efficient billing and to enable smart grid integration [6]. The used cloud solutions can also collect and store energy consumption data that can be used later for data analytics proposes for finding new trends in energy consumption and using them for further enhancements. Furthermore, some industrial and commercial products and services that are integrated with and benefit from the cloud are being offered to smart buildings. One example is Enterprise Energy Management Suite (EEM Suite) from Mckinstry [41]. The EEM Suite mainly offers a set of tools that can assist businesses in managing their multi-site energy and water usage and costs in a single Internet-based system. It is designed not only for industrial facilities but also can be used for other types of buildings. It involves meter data gathering, unusual pattern detection, energy demand prediction, and real-time monitoring.

Another product is ClockWorkers from KGS Buildings [42]. ClockWorkers provides a set of automated analytic libraries to schedule prioritized energy efficiency and maintenance activities for facility maintenance unit and service providers and can deal with multiple sites. It utilizes information collected in real-time from energy meters, equipment, and the environment to provide mechanisms to detect unusual patterns, energy efficiency information, and equipment efficiency levels. Another industrial solution is the Energy Management System from Powertech [43]. It collects information from energy meters, equipment, and facility environment to help in providing energy cost allocations and real-time controls. In addition, EnerNOC introduced Demand Manager, software-as-a-service (SaaS) that offers power utilities and retailers with tools to manage their demand response programs [44]. This includes real-time load monitoring, measurement, energy-efficient advisory, energy cost allocation, and energy demand prediction.

5.9 Conclusion

Smart buildings energy management is an important component to enhance the operations and efficiency of energy systems. As stand alone, they can offer limited benefits and capabilities. The cloud has proven to be an effective model to provide large storage space and advanced software services that can support various types of systems. Combining the benefits of both through the use of an energy cloud is a logical step towards achieving better results and enhancing overall energy efficiency of smart buildings. The energy cloud can provide many services and benefits for energy management systems in smart buildings. These benefits include providing better energy efficiency, better fault detection and diagnosis, and better feasibility for building energy management systems. In addition, it can support research, analysis, and strategic planning. However, achieving the full benefits of the energy cloud involves solving several challenges including security and privacy, heterogeneity, data management, and maintainability. There are some research efforts to provide some solutions for some of the perceived challenges. In addition, some companies started to utilize the concept of the energy cloud to provide cloud-based energy services to their customers. However, each of these services covers one or a few features and leaves out many other possible features. As a result, more research and industrial efforts are still required to provide comprehensive solutions for achieving ideal energy efficiency in smart buildings.

References

1. Architecture 2030. 2030 challenge for products: critical points. 2011.
2. Communication from the European Commission. Energy efficiency: delivering the 20% target (772). 2008.
3. Dounis AI, Caraiscos C. Advanced control systems engineering for energy and comfort management in a building environment—a review. Renew Sust Energ Rev. 2009;13(6): 1246–61.

4. Armbrust M et al. Above the clouds: a berkeley view of cloud computing, electrical engineering and computer sciences. University of California at Berkeley, Technical Report No. UCB/EECS-2009-28, http://www.eecs.berkeley.edu/Pubs/TechRpts/2009/EECS-2009-28.html. 10 Feb 2009.
5. Hanna EM, Mohamed N, Al-Jaroodi J. The cloud: requirements for a better service. In: 2012 12th IEEE/ACM international symposium on cluster, cloud and grid computing (CCGrid). IEEE; 2012. p. 787–92.
6. Alahakoon D, Yu X. Smart electricity meter data intelligence for future energy systems: a survey. IEEE Trans Ind Inf. 2016;12(1):425–36.
7. Schor L, Sommer P, Wattenhofer R. Towards a zero-configuration wireless sensor network architecture for smart buildings. In: Proceedings of the first ACM workshop on embedded sensing systems for energy-efficiency in buildings. ACM; 2009. p. 31–6.
8. Huang W, Lam HN. Using genetic algorithms to optimize controller parameters for HVAC systems. Energ Buildings. 1997;26(3):277–82.
9. Kanarachos A, Geramanis K. Multivariable control of single zone hydronic heating systems with neural networks. Energy Convers Manag. 1998;39(13):1317–36.
10. Yao Y, Lian Z, Hou Z, Zhou X. Optimal operation of a large cooling system based on an empirical model. Appl Therm Eng. 2004;24(16):2303–21.
11. Alcalá R, Casillas J, Cordón O, González A, Herrera F. A genetic rule weighting and selection process for fuzzy control of heating, ventilating and air conditioning systems. Eng Appl Artif Intell. 2005;18(3):279–96.
12. Fong KF, Hanby VI, Chow TT. HVAC system optimization for energy management by evolutionary programming. Energ Buildings. 2006;38(3):220–31.
13. Kolokotsa D, Stavrakakis GS, Kalaitzakis K, Agoris D. Genetic algorithms optimized fuzzy controller for the indoor environmental management in buildings implemented using PLC and local operating networks. Eng Appl Artif Intell. 2002;15(5):417–28.
14. Clark G, Mehta P. Artificial intelligence and networking in integrated building management systems. Autom Constr. 1997;6(5):481–98.
15. Doukas H, Patlitzianas KD, Iatropoulos K, Psarras J. Intelligent building energy management system using rule sets. Build Environ. 2007;42(10):3562–9.
16. Zhao P, Suryanarayanan S, Simões MG. An energy management system for building structures using a multi-agent decision-making control methodology. IEEE Trans Ind Appl. 2013;49(1):322–30.
17. Wang L, Wang Z, Yang R. Intelligent multiagent control system for energy and comfort management in smart and sustainable buildings. IEEE Trans Smart Grid. 2012;3(2):605–17.
18. Figueiredo J, Costa JS d. A SCADA system for energy management in intelligent buildings. Energ Buildings. 2012;49:85–98.
19. Kang SJ, Park J, Oh KY, Noh JG, Park H. Scheduling-based real time energy flow control strategy for building energy management system. Energ Buildings. 2014;75:239–48.
20. Missaoui R, Joumaa H, Ploix S, Bacha S. Managing energy smart homes according to energy prices: analysis of a building energy management system. Energ Buildings. 2014;71:155–67.
21. Katipamula S, Brambley MR. Review article: methods for fault detection, diagnostics, and prognostics for building systems—a review, Part I. HVAC&R Res. 2005;11:3–25.
22. Park K, Kim Y, Kim S, Kim K, Lee W, Park H. Building energy management system based on smart grid. In: 33rd IEEE international telecommunications energy conference (INTELEC). IEEE; October 2011. p. 1–4.
23. Badger L, Grance T, Patt-Corner R, Voas J. DRAFT cloud computing synopsis and recommendations. NIST Spec Publ. 2011;800:146.
24. Atzori L, Iera A, Morabito G. The internet of things: a survey. Comput Netw. 2010;54(15):2787–805.
25. Parwekar P. From internet of things towards cloud of things. In: 2nd international conference on computer and communication technology (ICCCT). IEEE; 2011. p. 329–33.
26. Mohamed N, Lazarova-Molnar S, Al-Jaroodi J. CE-BEMS: a cloud-enabled building energy management system. In: Proceeding of 3rd MEC international conference on big data and smart city (ICBDSC 2016). IEEE: Muscat, Oman; 15–16 Mar 2016.

27. Lazarova-Molnar S, Mohamed N. Towards collaborative data analytics for smart buildings. In: Proceedings of 8th iCatse conference on information science and applications (ICISA 2017). Springer: Macau, China; 2017.
28. Liu L, Kuo SM, Zhou M. Virtual sensing techniques and their applications. In: IEEE international conference on networking, sensing and control (ICNSC'09), March 2009. p. 31–6.
29. Lazarova-Molnar S, Shaker HR, Mohamed N. Reliability of cyber physical systems with focus on building management systems. In: Proceeding of IEEE International performance computing and communications conference (IPCCC 2016), Las Vegas, Nevada, USA; 9–11 Dec 2016.
30. Xiao Z, Xiao Y. Security and privacy in cloud computing. IEEE Commun Surv Tut. 2013;15(2):843–59.
31. Curry E, O'Donnell J, Corry E, Hasan S, Keane M, O'Riain S. Linking building data in the cloud: Integrating cross-domain building data using linked data. Adv Eng Inform. 2013;27(2):206–19.
32. Hong I, Byun J, Park S. Cloud computing-based building energy management system with zigbee sensor network. In: 2012 sixth international conference on innovative mobile and internet services in ubiquitous computing (IMIS). IEEE; 2012. p. 547–51.
33. Javed A, Larijani H, Ahmadinia A, Emmanuel R, Mannion M, Gibson D. Design and implementation of cloud enabled random neural network based decentralized smart controller with intelligent sensor nodes for HVAC. IEEE Internet Things J. 2016;4(2):393–403.
34. Bruton K, Raftery P, O'Donovan P, Aughney N, Keane MM, O'Sullivan DTJ. Development and alpha testing of a cloud based automated fault detection and diagnosis tool for Air Handling Units. Automat Constr. 2014;39:70–83.
35. Lazarova-Molnar S, Mohamed N. A framework for collaborative cloud-based fault detection and diagnosis in smart buildings. In: Proceedings of 7th International conference on modeling, simulation and applied optimization. IEEE: Sharjah, UAE; 4–6 Apr 2017.
36. Mohamed N, Lazarova-Molnar S, Al-Jaroodi J. SBDaaS: smart building diagnostics as a service on the cloud. In: Proceedings 2nd International conference on intelligent green building and smart grid (IGBSG 2016). IEEE: Prague, Czech Republic; 27–29 Jun 2016.
37. Al Faruque MA, Vatanparvar K. Energy management-as-a-service over fog computing platform. IEEE Internet Things J. 2016;3(2):161–9.
38. Sequeira H, Carreira P, Goldschmidt T, Vorst P. Energy cloud: real-time cloud-native energy management system to monitor and analyze energy consumption in multiple industrial sites. In: Proceedings of the 2014 IEEE/ACM 7th international conference on utility and cloud computing. IEEE Computer Society; 2014. p. 529–34.
39. Motegi N, Piette MA, Kinney S, Herter K. Web-based energy information systems for energy management and demand response in commercial buildings. Lawrence Berkeley National Laboratory; 2003.
40. Papantoniou S, Kolokotsa D, Kalaitzakis K. Building optimization and control algorithms implemented in existing BEMS using a web based energy management and control system. Energ Buildings. 2014;98:45–55.
41. EEM Suite Website, http://www.mckinstryeem.com/, viewed on 6 Mar 2017.
42. ClockWorkers WebSite: http://www.kgsbuildings.com/clockworks, viewd on 6 Mar 2017.
43. Powertech Website: http://www.power-tech.com.tw/_english/01_energy/02_list.php?pid=11, viewed on 6 Mar 2017.
44. EnerNOC Utility Demand-Response Product Website: https://www.enernoc.com/products/products-utilities/demand-response, viewed 6 Mar 2017.

Chapter 6
Dynamic Virtual Machine Consolidation Algorithms for Energy-Efficient Cloud Resource Management: A Review

Md Anit Khan, Andrew Paplinski, Abdul Malik Khan, Manzur Murshed, and Rajkumar Buyya

6.1 Introduction

As envisioned by Leonard Kleinrock [1], Cloud computing has transformed the dream of "computing as a utility" into reality, so much so it has turned out as the latest computing paradigm [2]. Cloud computing is called as *Service-on-demand*, as Cloud Service Providers (CSPs) assure users about potentially unlimited amount of resources that can be chartered on demand. It is also known as *elastic computing*, since Cloud Service Users (CSUs) can dynamically scale, expand, or shrink their rented resources anytime and expect to pay for the exact tenure of resource usage under Service Level Agreements (SLA). Through such flexibilities and financial benefits, CSPs have been attracting millions of clients who are simultaneously sharing the underlying computing and storage resources that are collectively known as Cloud data centers.

However, as the number of CSUs is increasingly growing over time, the necessity of an automatic management system, referred to as Cloud Resource Management System (CRMS), has become imminent to manage the multitude of service requests of millions of CSUs in automatic fashion. Furthermore, several research unlock the fact that Cloud Data Centers (CDCs) consume a lot of electricity which on average is twenty-five thousand times more than a household's energy consumption.

Md.A. Khan (✉) • A. Paplinski • A.M. Khan
Faculty of Information Technology, Monash University, Clayton, VIC, 3168, Australia
e-mail: makha23@student.monash.edu

M. Murshed
School of Information Technology, Faculty of Science, Federation University Australia, Churchill, VIC, 3842, Australia

R. Buyya
Cloud Computing and Distributed Systems (CLOUDS) Laboratory, School of Computing and Information Systems, The University of Melbourne, Parkville, VIC, Australia

W. Rivera (ed.), *Sustainable Cloud and Energy Services*,
DOI 10.1007/978-3-319-62238-5_6

Subsequently, more carbon dioxide (CO_2) emission is driven by these data centers. A study has revealed that energy consumption of Cloud causes more carbon emission to that of two countries, Netherlands and Argentina [3]. In [4], author has highlighted that worldwide energy usage of cloud data centers climbed up to 56% between 2005 and 2010, and was projected to be between 1.1 and 1.5% of the total electricity use in 2010 [5]. Moreover, approximate carbon emission by IT industry is 2% of the global emissions which is equivalent to the emissions of the aviation industry [6]. Consequently, many developed countries are getting more concerned these days to reduce carbon emission [7]. Beyond that, Koomey [8] projected that energy consumption of CDCs would remain to rise rapidly without energy-efficient resource management solutions being applied. As such, development of energy-efficient CRMS is extremely crucial.

VM Consolidation (VMC) is one such technique incorporated in CRMS to increase the energy-efficiency of Cloud. Hardware failure of existing Physical Machines (PMs) and addition of new PMs are continuous events in data center. Furthermore, resource requirement to accomplish the remaining tasks of existing service requests evolves with the course of time. Hence, as time progresses, remapping of remaining workload to currently available resources become inevitable to uphold the optimization of Cloud resource usage. The VM consolidation technique is applied to remap the remaining workloads to currently available resources which opts to migrate VMs into lesser number of active PMs, so that the PMs which would have no VM can be kept into sleep state. Since, energy consumption by the PM which is in sleep state is significantly lower than the energy consumption by the PM which is in active state, therefore, VM consolidation minimizes energy consumption of Cloud data center.

As portrayed in Fig. 6.1, before VMC is applied, VMs are scattered in multiple PMs. VMC migrates the VMs from lower utilized PMs to higher utilized PMs and thus consolidate VMs in lesser number of PMs than before. Meanwhile, the state of those PMs having no VMs can be changed from active state (i.e., turned on state) into a lower energy consuming state, such as sleep state and consequently, energy consumption can be minimized. The more number of VMs are placed in one single PM, the lesser becomes the overall energy consumption with the lesser number of active servers. Moreover, on account of compacting more number of VMs into fewer number of PMs, *resource utilization ratio* of the PM P_i, R_{P_i} (6.1) would become higher which in turn would increase the *mean resource utilization ratio of CDC*, $\overline{R}_{\text{CDC}}$ (6.2) where N is the total number of active hosts in CDC.

$$R_{P_i} = \frac{\text{Utilized Amount of Resource of } P_i}{\text{Total Amount Resource of } P_i} \tag{6.1}$$

$$\overline{R}_{\text{CDC}} = \frac{1}{N}\sum_{i=1}^{N} R_{P_i} \tag{6.2}$$

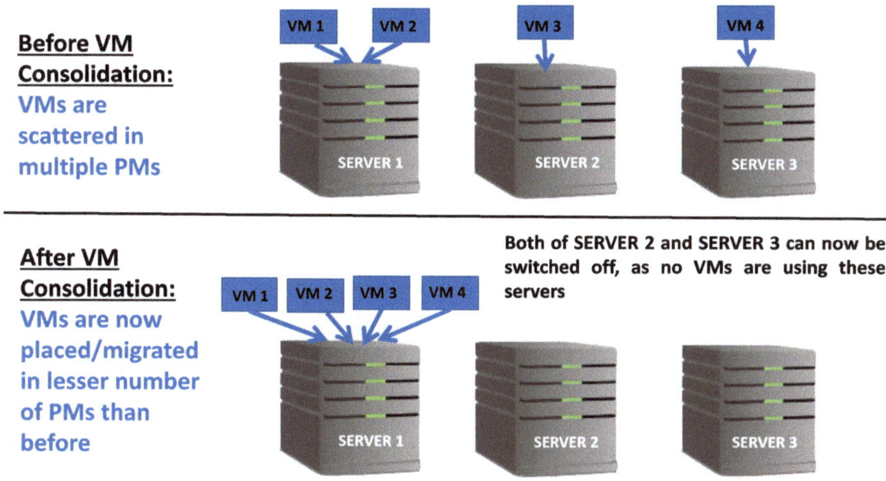

Fig. 6.1 VM consolidation

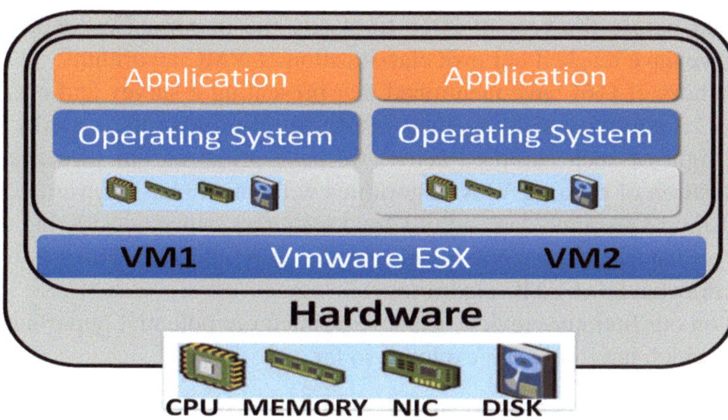

Fig. 6.2 Multiple VMs are hosted in a single physical machine [9]

However, as delineated in Fig. 6.2, VMs hosted in a PM share the underlying physical resources of that PM. Therefore, with the increased number of VMs sharing underlying resources of a single PM, waiting time for a VM prior receiving its required resources becomes higher. Thus, if more VMs are placed in a single PM, resource contention may arise which would lead towards poor QoS. Consequently, possibility of Service Level Agreement (SLA) violation arises.

To balance the tradeoff between QoS and energy-efficiency, it is extremely challenging to design such VMC algorithm which increases both resource utilization and energy-efficiency without compromising the QoS of running applications as agreed with respective CSUs during SLA. Recently, VMC has attracted interest of Cloud researchers, while designing efficient VMC algorithms is extremely challenging as it needs to be *scalable*, to the millions of VMs and PMs, as well as *robust*, such that the performance does not degrade with the fluctuation in resource demand of

VMs. In this paper, we have presented detail discussion on a wide range of VMC algorithms which complements three extremely important challenging aspects of Cloud: resource utilization maximization, energy consumption minimization, and Cloud profit maximization. In brief, our contributions are as follows:

- A group of surveys on VMC algorithms [10–20] have greatly assisted our research with VMC algorithms. However, VMC is an extremely popular research area and none of the available survey papers have presented discussion on VMC algorithms published in 2016 and onwards. Therefore, in this paper, we have primarily focused on most recent VMC algorithms published in 2016 and onwards which are not available in any other survey papers.
- We have presented critical review of different types of VMC algorithms which is missing in the available literature.
- There are several existing VMC algorithms proposed before the year of 2016 which strongly influenced subsequent researchers through introducing unique research directions. Those researches presented their own distinct techniques which are strong enough to be used as classification criterions. Therefore, we have also included elaborate discussion on such prominent VMC algorithms proposed before the year of 2016 to clarify the important concepts based on which we have reached our own classification of VMC algorithms.
- The authors of [16] have mentioned that presenting a survey and classification of VMC algorithms with an equal justice to all viewpoints is hardly possible. In the light of such admitted belief, we have proposed our own analysis and classification of existing VMC algorithms with an emphasis towards incorporation of future resource demand of Cloud resources, since considering the future resource demand is essential to prevent the SLA violation which is one of the major drawbacks of VMC algorithms.
- Based on our literature review, we have pointed out potential important research scopes which have not been explored so far.

The rest of the paper is organized as follows. In Sect. 6.2, we discuss the fundamental components of VMC. In Sect. 6.3, we present classification of VMC algorithms, followed by the extended classification and critical review of Dynamic VMC algorithms in Sect. 6.4. Next, in Sect. 6.5, we highlight the details of the notable contemporary VMC algorithms. Finally, we summarize our findings along with potential research directions in Sect. 6.6.

6.2 Fundamental Components of VMC

VMC algorithm is comprised of three core components [4, 21] which are as follows:

- *Source Host Selection*: First, among all the PMs, a set of PMs are selected from where VMs are migrated out. The *Source Host Selection* component takes all the PMs and VMs as input and selects one or more PMs as source PM(s) from where VMs would be migrated out.

- *VM Selection*: Second, one or more VM(s) are selected for migration from a source PM. The *VM Selection* component takes the PM as input which has been selected by *Source Host Selection* component and selects one or more VMs from that source PM for migration into a different PM.
- *Destination Host Selection/VM Placement*: Finally, the *Destination Host Selection/VM Placement* component selects a PM for each of the migrating VM which was selected by *VM Selection* component.

However, after a VM is created for the first time (i.e., for new VMs), the initial placement of that newly created VM can also be considered as VMC, if the corresponding destination PM is selected with an aim to minimize the total number of active PMs and increase the resource utilization, given that the hosting PM has adequate resources to fulfill the resource demand of the new VM. Hence, for new VMs, only *Destination Host Selection Algorithm/VM Placement Algorithm* does the VMC, as no source host selection and VM selection are needed for new VMs. A number of VM placement algorithms are available in the literature. In this paper, we have covered both VMC algorithms and VM placement algorithms, considering VM placement algorithms as VMC algorithms in case of new VMs. In the following section, we have presented the classification of VMC algorithms.

6.3 Classification of VMC Algorithms

In Fig. 6.3, we have delineated the classification of VMC algorithms. VMC algorithms can be broadly classified into two groups:

- *Dynamic VMC (DVMC) Algorithm*: In Cloud, received workloads are run in VMs, while these VMs accomplish the assigned workload through consuming the resources of the respective hosting PMs. With the advancement of time, progression of previously accepted workloads continues, while at the same time, new workloads keep being accepted by CSP. Furthermore, removal of some PMs due to hardware failure and addition of new PM also takes place. Thus, the overall workload with corresponding resource requirement and resource availability in the CDC keeps evolving over time. In DVMC algorithm, current VM-to-Server assignment is taken into consideration in the VMC process. Note that the workload or resource requirement of any VM and its location (i.e., its hosting PM) can be dynamic, as it changes with time. If the VMC algorithm consolidates VMs considering the dynamic (i.e., changing) workload and location of the VM (i.e., current VM-to-Server assignment), then it is called DVMC algorithm. In simple words, DVMC algorithms provide the solution of reallocation of existing VMs in lesser number of PMs such that the number of active PMs is minimized.
- *Static VMC (SVMC) Algorithm*: In contrast to DVMC algorithm, Static VMC (SVMC) algorithms, also referred to as consolidated VM placement algorithms, do not consider the current VM-to-Server assignment while choosing a new destination PM for any VM. In [22], the authors have mentioned that static

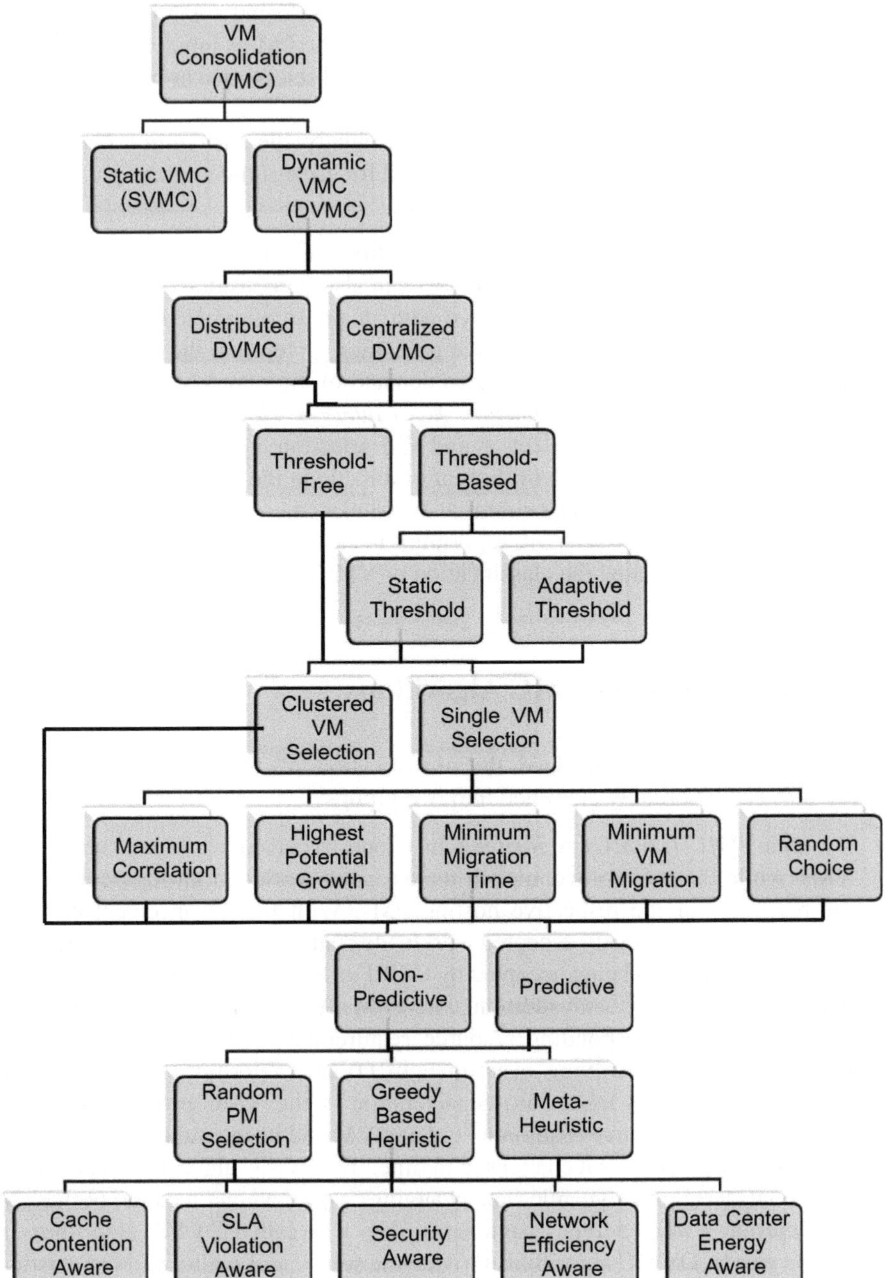

Fig. 6.3 Classification of DVMC algorithms

VMC algorithms work with a set of fully empty PMs and a set of VMs with specific resource requirement. In simple words, static VMC algorithm provides the solution of initial VM placement in minimum number of active PMs so that energy-efficiency and resource utilization of CDC increase. However, it does not provide the solution for reallocation of VMs in new PMs considering the current VM-to-Server assignment. Since, the dynamic (i.e., changing) load and placement of VMs are not considered, therefore, it is called as SVMC algorithm. [23–26] are examples of SVMC algorithm which do not consider the current VM-to-Server assignment while choosing a new destination PM for a VM.

Energy-efficiency of CDC would be hampered without the initial consolidated VM placement, as provided by SVMC algorithms. Besides, the energy-efficient initial placement keeps VMs consolidated in fewer PMs from the very beginning and consequently, the intermediate period before the awakening of the necessity to reallocate VMs can be prolonged. VMC has network overhead and it hampers QoS due to inherent service downtime. The prolonged period between initial VM placement and VMC or between two consecutive VMC reduces the overhead of VMC. However, the dynamic VM-to-Server assignment is not taken into consideration in SVMC algorithm. Therefore, the migration cost of a VM from its current hosting PM to its new destination PM is ignored. Consequently, SVMC algorithms are only applicable for initial placement of VMs or migrating VMs of one CDC into another CDC. As the time progresses, both workload and resource availability change in CDC. Therefore, apart from the initial consolidated VM placement, DVMC is one of the key techniques that uphold the energy-efficiency, resource usage optimization, and profit maximization of CSPs. As such, in this paper we have focused on DVMC algorithms. In the following section the classification of the DVMC algorithms has been presented.

6.4 Classification of DVMC Algorithms

DVMC problem focuses on run-time environments where VMs are active and already hosted by servers in the data center. Consolidation of such VMs is achieved by the VM live migration operations, where a running VM is relocated from its current host to another server while it is still running and providing service to its consumers [27]. DVMC algorithms can be classified into two groups:

- *Centralized DVMC Algorithm*: As proposed in [28–30], in centralized architecture, there is a single controller which has the information about present resource availability of all the PMs. The controller runs the centralized VMC algorithm which selects a destination PM for a migrating VM considering the resource availability of all the PMs.
- *Distributed DVMC Algorithm*: Instead of having a single controller which poses the information of present available resource availability of all the PMs and selects a destination PM for any migrating VM considering that information; in distributed architecture, PMs exchange information of their present resource

availability with their own neighbor PMs and thus, each PM has the resource availability information of its neighborhood PMs. If a PM wants to migrate out one of its VMs, it executes distributed VMC algorithm to select one of the neighbor PMs as the destination PM where the migrating VM would be hosted. Examples of distributed DVMC algorithms are [31, 32].

Majority of the DVMC algorithms found in the literature are centralized DVMC and only a few Distributed DVMC [31, 32] has been proposed. In [31], the authors have presented their distributed DVMC algorithm for a P2P network oriented CDC. According to [31, 32], the growing number of PMs becomes a bottleneck for centralized VMC at the time of selecting a destination for any migrating VM, since the asymptotic time complexity of the centralized DVMC algorithm is proportional to the number of PMs in the CDC, whereas the number of potential PMs to choose from a migrating VM is relatively small in distributed DVMC. Thus, distributed DVMC is more scalable for CDC with huge number of PMs. However, distributed DVMC has message passing overhead, as every PM must update its present resource availability to all of its neighbors. Every time a VM is migrated, both the sender PM and the destination PM must update their present resource availability status to all of their neighbors. Besides, a central monitoring system is indispensable in Cloud which monitors the accepted workload progression status and allocate/deallocate resources accordingly to accomplish the workload in time. Furthermore, at the time of accepting new workload, the overall resource availability status of CDC must be known, so that accurate decision can be made on whether the new workload would be possible to serve within deadline. Therefore, centralized DVMC can be implemented without adding any additional resources. Moreover, message passing as required by distributed DVMC algorithms increases network overhead, decreases network throughput, and increases network related energy consumption. Hence, centralized DVMC is more energy-efficient compare to distributed DVMC algorithms.

As discussed earlier in Sect. 6.2, first component of DVMC algorithm is to select the source PM. A DVMC algorithm can either randomly select a source PM from where one or more VM(s) are migrated out or VM(s) can be selected from over-utilized and under-utilized PMs. In the following section, we have presented our classification of DVMC algorithms based on different source PM selection techniques.

6.4.1 Classification of DVMC Algorithms Based on Different Source PM Selection Techniques

Based on the way source PMs are selected, DVMC algorithms can be classified into two groups:

- *Threshold-Based DVMC Algorithm*: Threshold-based DVMC algorithms use upper and lower threshold values to identify a PM as overloaded or underloaded, respectively, from the perspective of resource utilization ratio of the PM P_i, R_{P_i}

(6.1). The key point is that R_{P_i} is compared against the values of some thresholds which can be either static or adaptive (explained in detail later in this section). As presented in [33], if R_{P_i} goes past the upper utilization threshold value, then P_i is identified as overloaded or over-utilized PM and VMs are migrated out from P_i, until R_{P_i} becomes lower than the upper-threshold, since high R_{P_i} is a strong indicator of potential QoS degradation or SLA violation which is arisen because of the higher resource demand of the hosted VMs. Again, if R_{P_i} is smaller than the lower utilization threshold value, then P_i is identified as underloaded or under-utilized and potential destination PMs are looked for, where VMs of P_i can be migrated out so that P_i can be put in sleep state.

- *Threshold-Free DVMC Algorithm*: Unlike threshold-based DVMC algorithms, in threshold-free DVMC algorithms, resource utilization ratio of the PM P_i, R_{P_i} (6.1) is not compared against any threshold value to identify the PM as overloaded or underloaded. Instead, the source PMs are selected either randomly [34] or some functions are applied to favor PMs having either higher or lower R_{P_i} compare to those of other PMs. Examples of threshold-free DVMC algorithms are [25, 34, 35].

In [35], the authors have proposed two functions, one of which is used to select a source PM from where VMs would be migrated out and the other function is used to select destination PMs for those migrating VMs. The destination selection function favors PMs with neither higher nor lower R_{P_i}, while the source selection function favors PMs with lower R_{P_i}. It is noteworthy that no threshold value is used against which a PM's utilization is compared to identify it as over-utilized or under-utilized. Since, according to the proposed method, the PMs with higher R_{P_i} are not selected as source PMs to migrate out VMs; therefore, the QoS degradation or SLA violation due to increased resource demand of hosted VMs may take place.

The main difference between threshold-free and threshold-based DVMC algorithm is that threshold-based approach identifies a PM as overloaded or underloaded with respect to some thresholds, whereas threshold-free approaches select PMs either randomly or based on the lower or higher resource utilization ratio in comparison to those of rest of the PMs. As opposed to random source PM selection, the heuristic approach to always select the source PM with highest or lowest resource utilization ratio may not ensure the achieving of global best solution. Since, there is no random source PM selection policy in threshold-based DVMC algorithm; therefore, it may not provide the global best solution. Furthermore, compare to threshold-free approach, the number of VM migrations may become higher in threshold-based approach. To illustrate more, assume that overall workload in the CDC has become high. PMs would then experience high utilization as per threshold-based approach and would start migrating out their VMs, whereas threshold-free approach would consider the global picture of increase of workload in all PMs and may avoid VM migrations. However, VMC is a combinatorial optimization problem. In other words, it is a rearranging problem of VMs into different PMs, so that VMs can be placed in minimum number of PMs without violating the resource capacity of any PM and the total number of possible combinations is N^M, where N and M denote the total number of VMs and PMs, respectively. Hence, the solution search space would remain as humongous as N^M, if random selection policy is followed as per threshold-free approach. In contrast, although the solution may not be globally optimal, but selecting source PM based on upper or lower resource utilization threshold would confine the solution search space and would provide sub-optimal solution in faster time, since minimization of lower resource utilization certainly decreases the energy consumption.

From the literature review, threshold-based approaches are found as highly popular among researchers. Based on the type of used thresholds to identify a PM as overloaded or underloaded, threshold-based DVMC algorithms can be classified into two groups:

- *Static Threshold-Based DVMC Algorithm*: In static threshold-based DVMC algorithms, fixed values or predefined values are used as upper and lower thresholds to identify the PM as overloaded or underloaded. As the values of the thresholds do not change over time, therefore they are referred to as static thresholds. Examples of static threshold-based DVMC algorithms are [31, 33, 36]. In [36], the authors have used 100% CPU utilization as upper utilization threshold and 50% CPU utilization as lower utilization threshold. In other words, if the CPU utilization of a PM is found as 50%, then that PM is considered as lower utilized PM and VMs are migrated out from that PM into other PMs. Similarly, if the total resource demand of all the VMs hosted in a particular PM is found as higher than the CPU capacity of that PM, then that PM is considered as overloaded PM and VMs are migrated out from that PM into other PMs.
- *Adaptive Threshold-Based DVMC Algorithm*: In this case, the values of the thresholds based on which the PM is selected as overloaded or underloaded change dynamically as the resource utilization ratio the PM P_i, R_{P_i} (6.1) changes with time. In other words, the threshold value adapts with the change of resource utilization. Examples of adaptive threshold-based DVMC algorithms are [21, 29, 37, 38].

The authors of [21] are pioneers in proposing adaptive threshold-based DVMC algorithm, as they proposed a number of adaptive thresholds based on which a PM is detected as overloaded. One such adaptive threshold is referred to as Median Absolute Deviation (MAD). To illustrate more, let, $T = t_j \quad t_j$ denotes time j and $j \in \mathbb{N}$ where $\mathbb{N}$ is the set of positive integers and $R_{P_i}^{t_j}$ is the resource utilization ratio the PM P_i at time t_j. For each PM P_i, R_{P_i} (6.1) across different time (i.e., $R_{P_i}^{t_1}$, $R_{P_i}^{t_2}$, $R_{P_i}^{t_3}$, and so forth) would be recorded and then MAD is calculated using (6.3), while the upper utilization threshold T_u is calculated using (6.4), where $s \in \mathbb{R}^+$ a parameter which defines how strongly the system tolerates host overloads.

$$\text{MAD} = \text{Median}\left(\left|R_{P_i}^{t_j} - \text{Median}\left(R_{P_i}^{t_j}\right)\right|\right) \tag{6.3}$$

$$T_u = 1 - s \cdot \text{MAD} \tag{6.4}$$

The lower the value of s, the system is more tolerant to variation in resource utilization. If the current $R_{P_i}^{t_j}$ is found as greater than T_u, then P_i is considered as overloaded [4]. Note that the value of MAD (6.3) is not fixed or static, as $R_{P_i}^{t_j}$ changes with time and hence, T_u (6.4) also changes with the change in resource utilization.

As mentioned before, VMC algorithms aim to increase the resource utilization which helps to minimize the energy consumption. The high value of upper and lower utilization thresholds increases resource utilization and energy-efficiency. However, very high resource utilization or higher R_{P_i} (6.1) causes QoS degradation or potential SLA violation. Therefore, the higher is the upper utilization threshold, the higher is the SLA violation. Hence, there is a tradeoff between energy-efficiency and SLA violation, while the values of upper and lower utilization thresholds have great impact in controlling such tradeoff. Therefore, a balance is needed to be maintained between energy-efficiency and SLA violation through controlling the threshold values.

The key idea of using upper and lower static thresholds is to keep the resource utilization restricted into a certain range (i.e., in between upper and lower utilization threshold), so that a balance exists between energy-efficiency and SLA violation. However, the workload pattern as experienced by an application running inside of a VM changes over time. Besides, multiple VMs which are all hosted in a single PM may exhibit different workload pattern. Since, the threshold is static and it cannot be changed with the change of workload, therefore, SLA violation increases with the increase of workload.

Adaptive threshold-based DVMC algorithms partially mitigate this problem by changing the utilization threshold values with the variation in workload. For instance, as the workload grows in VMs, MAD (6.3) increases and the upper utilization threshold (6.4) becomes lower accordingly. Consequently, VMs are migrated out from P_i before R_{P_i} (6.1) reaches to a very high level and as a result, VMs of P_i do not suffer from degraded QoS due to high R_{P_i} (6.1). In general, compared to static threshold-based approach, adaptive threshold-based approach decreases SLA violation rate more. However, it provides less energy-efficiency than static threshold-based approach, since the lower is the upper and lower utilization threshold, the lower is the energy consumption minimization. Apart from that, adaptive-based approach causes more number of VM migration than static-based approach which increases both energy consumption and SLA violation.

Thus far, we have reviewed different types of threshold-based DVMC algorithms and previously in Sect. 4.1, we have presented our comparison between threshold-free DVMC algorithm and threshold-based DVMC algorithm. As highlighted in Sect. 6.2, one of the core components of VMC algorithm is VM selection. In the following section, we have presented our discussion on different DVMC algorithms with different VM selection policies.

6.4.2 Classification of DVMC Algorithms Based on VM Selection Policy

Once source PMs are selected, the following step of VMC is to select one or more VM(s) from source PM to migrate out. Based on the different VM selection policies used in different DVMC algorithms, the DVMC algorithms can be broadly classified into two groups:

- *Clustered VM Selection (CVS)*: Multi-layered applications comprised with load balancer(s), application server(s), and database server(s) are hosted in Cloud, while individual VM is assigned with the functionalities of each layer. To explain more, database server instance(s) are deployed in one or more VMs, while load balancer(s) are run in separate VMs and application server(s) are run in different VM(s). All these VMs which are under one application communicate with each other and the performance of the application becomes immensely hampered if these VMs are not hosted in nearby PMs. As such, instead of migrating a single VM, such group of VMs of an application, also known as Clustered VMs, are considered for migration. Prominent examples of such clustered VM selection algorithms are [24, 39].
- *Single VM Selection (SVS)*: Unlike CVS, SVS algorithms select a single VM to migrate out. Different single VM selection strategies as found in the literature are mentioned in the following:
  - *Random Choice (RC)*: Among all the VMs residing in the source PMs, a VM is randomly selected [33, 40]. Random VM Selection can select a VM in O(1) time which is faster than rest of the approaches.
  - *Minimization of VM Migration (MVM)*: Minimum number of VMs are migrated to make the current resource utilization of a PM lower than the upper utilization threshold. MVM algorithm as proposed by [33] first sorts the VMs in descending order with respect to CPU demand and then selects the VM that satisfies the two criterions: First, the VM's CPU utilization should be higher than the difference between the host's present overall CPU utilization and the upper utilization threshold; Second, that VM is selected for which the difference between the upper-threshold and the new utilization is the minimum compare to the values provided by all the VMs. If no such VM is found, then the VM with the highest utilization is selected and the process is repeated until the new utilization becomes lower than the upper utilization threshold.
  - *High Potential Growth (HPG)*: The VM with lowest ratio of actual resource usage to its initial claimed resource demand is selected [33]. Asymptotic running time of the algorithm is O(n).
  - *Minimization of Migration Time (MMT)*: The VM which requires minimum time to complete the migration is selected for migration, while the migration time is estimated as the amount of RAM utilized by a VM divided by the spare network bandwidth available for the hosting PM [21]. Asymptotic running time of the algorithm is O(n).

- *Maximum Correlation (MC)*: VM that has the highest correlation of the resource utilization with other VMs is selected [4]. Multiple Correlation Coefficient as proposed by [41] is used to determine the correlation between the resource utilization of VMs.

The RC may help to find the globally optimal solution, if source PM is selected randomly using threshold-free approach as discussed earlier in Sect. 4.1. Furthermore, RC shows the fastest run time among all the above-mentioned VM selection algorithms. However, if the solution space is confided prior, such as source PM is selected using the heuristic that the PM with highest or lowest resource utilization would be the source PM and after then that, a VM is randomly chosen from that PM, then RC cannot provide the global optimal solution. As oppose to that, after confining the solution search space by selecting source PM based on highest or lowest resource utilization ratio, rest of the heuristics such as MVM, HPG, MMT, and MC are more likely to decrease the energy consumption and SLA violations compare to RC; since, heuristics like MVM, HPG, MMT, and MC certainly provide the local best solution, whereas RC probabilistically chooses a solution which may not be locally optimal. To illustrate more, VMs experience degraded QoS during the period of migration. Therefore, selecting the VM which would take the least migration time (i.e., MMT) would certainly assist in keeping the SLA violation lower [42]. In contrast, RC may choose a VM with higher migration time. Consequently, SLA violation rate would be higher for RC.

MVM minimizes the number of VM migrations with minimal decreasing of resource utilization ratio of the PM P_i, R_{P_i}, (6.1). As a result, *mean resource utilization ratio of CDC* R_{CDC} (6.2) would remain as higher compare to rest of the above-mentioned algorithms and thus it turns out to be more energy-efficient. However, higher R_{P_i} (6.1) may cause degraded QoS and more SLA violation. Hence, MVM shows lower SLA violation than MTM.

Another critical issue with MVM is that VMs need to be sorted first with respect to resource utilization, as otherwise the asymptotic running time is exponential. However, it is not possible to sort the VMs with respect to resource demand, since a VM has three different types of resource demand, such as CPU demand, memory demand, and network bandwidth demand which are not related. For instance, a VM may have high CPU demand and low network bandwidth demand, whereas another VM may have low CPU demand and high network bandwidth demand. Therefore, it is not possible to sort VMs based on VM resource demand, since one distinctive feature of CDC is location transparency [43] which arises from the fact that a VM can be placed in any of the PMs. Hence, a PM may have VMs with varied resource demand with respect to different resource types [44].

Because of such diverse resource utilization value of a VM across various types of resources, it is not possible to select a VM with highest potential growth ratio (i.e., ratio of actual resource usage to a VM's initial claimed resource demand) among all the VMs across all resource types. Consequently, HPG is only possible to be implemented considering one resource type, such as CPU or memory or network bandwidth and thus, it does not ensure the minimization of energy-efficiency or

SLA violation. Similar issue exists with MC algorithm. In contrast, since MTM primarily selects VM based on memory size and hence, it is free from such issue.

Thus far, we have analyzed different DVMC algorithms with different VM selection policies. Another critical distinguishing aspect among DVMC algorithms is that whether estimated future resource demand has been considered in the VMC process or not, as we have presented our discussion about it in the following section.

6.4.3 Classification of DVMC Algorithms based on Consideration of Estimated Future Resource

From the literature, it can be viewed that consideration of estimated future workload in a PM is commonplace in a wide range of DVMC algorithms. However, there are still numerous DVMC algorithms which make the consolidation decision based on the current resource utilization of PMs instead of the estimated future resource utilization. Consideration of future can create a significant difference on the performance of VMC algorithm, compare to those VMC algorithms which takes the decision based on the current resource utilization. Hence, in this section, we have reviewed both types of VMC algorithms from that perspective.

- *Non-Predictive Dynamic VMC Algorithm (NPDVMC)*: Instead of considering the estimated future resource utilization of the PM, NPDVMC algorithms consider the current aggregated resource demand of VMs. Note that the aggregated resource demand of hosted VMs in the PM is equal to the resource utilization of that PM. VM migration decisions are taken when the current resource utilization of the PM P_i, R_{P_i} (6.1) becomes very high or very low so that SLA violation can be avoided or energy consumption can be minimized.

  One such example of NPDVMC algorithm is [28], where source and destination PMs for consolidation of VMs are selected based on the current resource utilization status of the PM. If the current R_{P_i} is found as equal or greater than 90%, then P_i is considered as overloaded or over-utilized and VMs are migrated out from P_i. Again, if current R_{P_i} is found as equal or lower than 10%, then P_i is considered as overloaded or over-utilized and VMs are migrated out from P_i to place in new PMs. Other prominent non-predictive VMC algorithms are [24, 25, 28–30, 34, 45].
- *Predictive Dynamic VMC Algorithm (PDVMC)*: On the contrary, PDVMC algorithms take the decision to migrate VMs from one PM to another PM considering the estimated future resource demand of VMs instead of current resource demand. Examples of PDVMC algorithms are [36, 46].
- In [36], both current and future resource utilization of the PM are considered while making the consolidation decision. Linear regression [38] is used to generate an estimated future resource utilization of a PM from analyzing its past resource utilization statistics. If the current resource utilization of the PM is found as higher than the upper-utilization threshold (explained previously in

Sect. 4.1), then that PM is identified as *overloaded*. Furthermore, although the current resource utilization of the PM is found as lower than the upper-utilization threshold, yet its estimated future resource utilization is found as higher than the upper-utilization threshold, then that PM is identified as *predicted overloaded*. Both *overloaded* and *predicted overloaded* PMs are elected as source PMs from where one or more VM(s) are selected to migrate out into new PMs. Again, both *overloaded* and *predicted overloaded* PMs are excluded from the list of potential destination PMs where migrating VMs would be placed.

– Consolidation of VMs considering the estimated workload is a more proactive approach than NPDVMC, as VMs are migrated out from those PMs which are predicted to be overloaded in future. The aim of such proactive approach is to move VMs out prior QoS degradation or SLA violation takes place. Consequently, compare to NPDVMC algorithms, PDVMC algorithms will display lower SLA violations due to less occurrences of resource contention. However, because of migrating more VMs out of the higher utilized hosts than NPDVMC, the *mean resource utilization ratio of CDC* $\overline{R}_{\text{CDC}}$ (6.2) would become lower and thus, total number of inactive PMs may become less for PDVMC. Hence, PDVMC would display lower energy consumption minimization than that of NPDVMC.

Another challenging aspect of PDVMC is that PDVMC relies on prediction techniques to estimate the future resource utilization of PMs. Predictive techniques are based on the correlation between the past history of the system behavior and its near future [4]. The efficiency of prediction-based techniques greatly depends on the actual correlation between past and future events and the quality of adjustment with a specific workload type. In Cloud environment, different VMs are hosted in a single PM, while these VMs are expected to exhibit different behavioral pattern from each other in terms of resource demand. Consequently, no single prediction technique would be a perfect fit for all PMs. A non-ideal prediction causes over- or underprediction which lead towards either inefficient resource usage or more SLA violation.

Until now we have reviewed diverse approaches to select source PMs and VMs, as we have also analyzed both prediction-based and non-prediction-based DVMC algorithms. One of the core components of DVMC algorithms is destination PM selection where migrating VMs are placed. This is also referred as *VM placement problem*. In the following section, we have presented our discussion on diverse approaches to select destination PMs for migrating VMs as incorporated in different DVMC algorithm.

6.4.4 Classification of DVMC Algorithms based on Destination PM Selection Strategies

Destination PM selection strategy plays a gigantic role in increasing the energy-efficiency of CDC. The aim of destination PM selection is to select such new PMs for migrating VMs so that the total number of active PMs becomes minimum without violating the resource constraint of any PM. However, destination PM selection component itself is a NP-hard problem and hence a number of heuristic as well as meta-heuristic algorithms have been proposed in the literature. Based on different destination PM selection strategies DVMC algorithms can be classified into three groups:

- *Random PM Selection (RPS)*: As proposed in [24], the destination PM is randomly selected from the suitable PMs for a given VM. The asymptotic running time of FF is O(n), where n is the total number of VMs.
- *Greedy Heuristic*: Greedy Heuristic algorithms are most widespread in the literature to select the destination PM for migrating VMs. Several popular heuristic-based algorithms are as follows:
  - *Random Choice (RC)*: As discussed in [47], RC algorithm randomly chooses a destination PM for a migrating VM which is already turned on. If no suitable PM can be found to host the migrating/target VM, then a new PM which was in sleep state is turned on to accommodate the target VM.
  - *First Fit (FF)*: In FF [47], PMs are ordered in a sequence and for each VM, the first available PM from the ordered list of PMs is selected. In other words, for every single VM, the searching of destination PM always starts from the first PM. If the first PM cannot accommodate a VM, then the second PM is checked and if the second PM cannot accommodate it, then the third PM is checked, as the searching continues to the next PM while always following the initial order of PMs until a suitable destination PM with adequate resource capacity is found. Since, a VM may have larger resource demand than the available remaining resource of a PM, therefore, the asymptotic running time of FF is O(nm), where n is the total number of VMs and m denotes the total number of PMs.
  - *First Fit Decreasing (FFD)*: FFD is same as FF, except the VMs are first sorted in the decreasing order of their resource demand. Then the destination PM for the first VM with highest resource demand is first searched using FF algorithm, as the searching continues for the VM with second highest resource demand, and so on. The asymptotic running time of FF is O(nlogn + nm), where n is the total number of VMs and m denotes the total number of PMs. Note that O(nlogn) is running time of sorting algorithm.
  - *Next Fit (NF)/Round Robin (RR)*: Like FF, NF also performs a sequential search, except it starts from the last server selected in the previous placement [39]. To explain more, if the last VM was placed in the second PM, then checking will start from the second PM for the following VM placement, and

so on [47], whereas in FF and FFD the checking would have always started from the first PM for any VM. NF is also referred to as *Round Robin (RR)*. The asymptotic running time of NF is same as FF.
- *Best Fit (BF)*: In BF, the PM with the minimum residual resource is selected as its destination PM [46]. The residual resource of the PM is the difference between the total resource capacity of that PM and the aggregated resource demand of the hosted VMs in it along with the resource demand of the target VM for which destination PM is under search. If PMs are first sorted based on resource utilization ratio (6.1), then running time of BF would be identical to that of FFD. However, if no sorting is applied, then the running time of BF would be $O(nm^2)$.
- *Best Fit Decreasing (BFD)*: VMs are first sorted in the decreasing order based on their resource demand. Then the destination PM for the first VM with highest resource demand is first searched using BF algorithm, as the searching continues for the VM with second highest resource demand, and so on. The asymptotic running time of BFD is same as FFD.
- *Power Aware Best Fit Decreasing (PABFD)*: PABFD proposed by [33] is a modified version of BFD, as the VMs are first sorted in decreasing order based on their CPU demand and then the destination PM is selected with the least power increase compare to all the suitable PMs which could host the target VM. The asymptotic running time of PABFD is same as FFD.

RPS is most time efficient, but least optimal compare to rest of the above-mentioned VM selection strategies from the perspective of energy-efficiency, since it does not ensure to first choose a suitable PM from the set of currently turned on PMs, so that unnecessary waking up of PMs which are currently in sleep state can be avoided.

Unlike RPS, the similarity among all the above-mentioned heuristic-based destination PM selection algorithms is that all these algorithms first attempt to select a PM from one of the PMs which are already in turned on state, so that energy consumption can be minimized. However, the difference exists in their selection strategy to select a PM for a VM among all the existing turned on PMs. Although RC first tries to select a PM from the existing turned on PMs, yet, because of its random PM selection nature, it may cause sparse VM placement or VMs may be found as more scattered compare to those of FF, FFD, BF, and BFD. Hence, for many VMs, suitable PMs would not be found which would result in turning those PMs on which were in sleep state or in switched off state. Consequently, energy consumption would be higher for RC compare to rest of the heuristics.

The rationale of BF heuristic is that placing VMs on the PM with the least remaining available resource would provide other turned on PMs with large remaining available resources which can be used to support future larger VMs, while this strategy would concomitantly increase the *mean resource utilization ratio of CDC*, $\overline{R}_{\text{CDC}}$ (6.2). Experimental results of [47] suggest that energy consumption is the highest for RC, while it is lowest for BF and BFD, as BF and BFD packs the VMs more tightly compare to RC, FF, and FFD.

Difference between BFD and PABFD is that BFD will select the smallest PM among all the suitable PMs for the first VM in terms of total resource capacity of PMs and then consolidating more VMs into it, whereas PABFD will initially select the most power-efficient PM among all the suitable PMs for the first VM and then consolidating more VMs into it. PABFD focuses on utilizing power-efficient PMs more which certainly has an impact on increasing the energy-efficiency of CDC. On the contrary, BFD leads towards utilizing the smallest PMs first and leaving larger PMs for future, while ignoring to ensure the usage of power-efficient PMs. With the increased number of VMs, larger PMs have to be turned on eventually. Hence, as opposed to BFD, since, PABFD ensures the more usage of power-efficient PMs, therefore PABFD is more energy-efficient compare to BFD.

- *Meta-heuristic*: Greedy Heuristic algorithms may become stuck with local minima or local maxima. Therefore, several meta-heuristic-based destination PM selection algorithms have been proposed in the literature which are discussed in the following:
  - *Evolutionary Algorithm*: The general steps of evolutionary algorithm are as follows:

    At each step (generation), the algorithm starts working with a population comprised of a range of solutions (members), while different evolutionary-based VMC algorithms use different heuristics to generate the initial solution.

    Next, a few members are first selected, also called parents from the generation to produce new solutions (children). Different evolutionary algorithms propose different methods to select parents from the population. One common method to select parents is to check value(s) of objective function(s) for each of the member and then select the members with higher values. The optimization functions are called as Pareto set and the values of the Pareto set achieved by the members are called as Pareto front [39]. For VMC, one prevalent object function is to maximize the number of physical servers with no VMs running in it or minimize the number of active physical servers.

    In order to produce a new solution (child), different parts collected from different parents are combined together. This technique is called mutation which takes place with a certain probability. The objective of mutation is the faster production of more optimized solutions.

    Furthermore, after mutation, swapping or interchanging (crossover) among different parts of a child takes place with a certain probability. The goal of crossover is to complement the faster creation of new children that are more optimized. In the context of VM placement or VMC, one widespread crossover technique is to interchange the hosts between two VMs [39].

    Through mutation and crossover, children are generated from parents which are added in the population.

    The entire process is repeated or more generations are run, until no improvements are found from consecutive repetitions of the algorithm.

Finally, a solution is chosen from the Pareto front based on an objective function.

- *Ant Colony Optimization (ACO)*: In ACO, a virtual ant selects a PM for a VM through considering two factors: Heuristic and Pheromone. Ants can either work sequentially or in parallel while constructing their own solutions. Each ant can either follow its own heuristic or all the ants can follow a common heuristic. The heuristic η is the key which guides to construct an optimal solution in aligned with the optimization function. Based on the diverse objective functions as presented in different ACO-based VM placement or VMC algorithms, the proposed η varies from one ACO-based VM placement or VMC to another. We have discussed about different heuristics previously. One common η found in a number of ACO-based VM placement or VMC is BFD [36]. Apart from η, the *Pheromone* plays a critical role in constructing an ant's solution which guides ants to find diverse solutions through exploring the search space. One key distinguishing aspect which makes ACO meta-heuristic different from heuristic-based algorithms is that some probability exists for an ant to choose the PM which is not optimal from the perspective of heuristic and thus stagnation into local minima or local maxima is avoided.
- *Simulated Annealing*: The authors of [35] have proposed a simulated annealing meta-heuristic-based VMC algorithm. In perturbation phase, instead of randomly choosing source or destination PMs, solutions are generated by selecting source PMs with lower utilization ratio and destination PMs with neither very high utilization ratio nor very low utilization ratio. Thus, VMs are consolidated in lesser number of active PMs. However, in order to avoid stagnation in local minima or local maxima, exploration is adopted through accepting solution that is even less optimal than the optimal solution found so far.

The aim of VMC is to minimize the energy-efficiency, as VMC minimizes energy consumption by placing more VMs in a single PM. The higher number of VMs is placed in a single PM, the higher is the minimization of energy consumption considering the energy spent after PMs. However, the higher number of VMs is placed in a single PM, the higher is the probability of QoS degradation or SLA violation. Hence, minimization of energy consumption and minimization of SLA violation are two confronting goals. While most researchers have focused to maintain a balance between minimization of PMs' energy consumption and SLA violation, some researchers have considered other aspects too, such as security, energy consumption by network, network throughput, and so forth. In the following section, classification of VMC algorithms based on their objectives has been presented.

6.4.5 Classification of DVMC Algorithms based on Different Objectives

The different DVMC algorithms with diverse objectives as observed in the literature are as follows:

- *SLA Violation Aware*: Since VMs share the underlying physical resources of their hosting PM such as CPU, RAM, and network bandwidth, therefore, the waiting time to receive the required resources for each VM increases with the increase of number of VMs in a single PM. Besides, the services which the VM is providing to its users' needs to be suspended temporarily at the time of migrating that VM from one PM to another PM. Hence, VMC causes SLA violation. Many VMC algorithms focus to minimize such SLA violation by limiting the number of VM migrations as well as minimize resource oversubscription and thus decrease SLA violation. [24] is an example of SLA violation aware VMC algorithms.
- *Security Aware*: Cloud is a multi-tenant environment, where VMs of different clients are hosted in same PM, while these VMs also share the underlying physical resources. Hence, security is one of the major challenging aspects in Cloud. In [28], the authors have proposed a security-based DVMC algorithm.
- *Network Efficiency Aware*: Such algorithms, for instance, [45] aim to uphold the network efficiency through considering different network related aspects, such as minimization of network energy consumption, and reduction of network congestion. [45, 48] are examples of network efficiency aware VMC algorithm.
- *Data Center Energy Aware*: CDC is the physical backbone of any Cloud-based services. One big challenging aspect of any data center is that an appropriate temperature must be always maintained, as the challenge escalates with the increase of the volume of data center along with the growth of number of PMs, network devices, and so forth. Cooling of CDC is extremely crucial to ensure the smooth and continuous functioning of PMs, routers, switches, and so forth. However, energy spending after cooling of CDC is very high and such energy requirement rises with the increase of quantity of VMs as well as with the increase of VMs' resource demand. Therefore, researchers have presented VMC algorithms which consolidate VMs in such a way that energy spending after cooling the data center can be minimized. [34] is an example of such VMC algorithm which minimizes the energy related to data center cooling.
- *Cache Contention Aware*: Cache contention refers to the situation that a VM may experience extra cache misses, as other VMs co-located on the same CPU fetch their own data into the Lowest Level Cache (LLC), which forces to evict the VM's data from the LLC and later fetching back that VM's data into the LLC again causes cache misses to other co-located VMs. In order to minimize cache misses due to VMC, [26] has proposed a cache contention aware VMC algorithm that considers the expected cache misses at the time of destination PM selection for a migrating VM.

Until now, we have reviewed different types of VMC algorithms, as we have highlighted the comparison with strengths and weakness of each of those techniques. We have delineated the classification of VMC algorithms in Sects. 6.3 and 6.4. In order to present further details of contemporary VMC algorithms, we have discussed about different aspects of recent VMC algorithms in the following section.

6.5 Detailed Analysis of Contemporary VMC Algorithms

Table 6.1 illustrates the most significant aspects of the notable recent research works on VMC as found in the published materials. The description of the attributes which are being considered to review the existing VMC algorithms is as follows:

- Research Project: Name of the reach project.
- Type of VMC: Whether the VMC is SVMC or DVMC.
- VMC Decision Process: If destination PM selection decision is taken centrally (i.e., centralized) or a source PM itself chooses another destination PM where the migrating VM will be placed (i.e., distributed).
- Resource Considered: Which resources (i.e., CPU, memory, network bandwidth, and so forth) are considered in the VMC algorithm.
- Source PM Selection Strategy: What the source PM selection strategy is and whether any threshold has been applied to select source PMs.
- VM Selection Criteria: What VM selection algorithm has been used.
- Application of Prediction Technique: Whether any prediction technique has been incorporated in the proposed system to predict the future resource utilization of PMs.
- Destination PM Selection Strategy: What algorithm has been used to select the destination PM for the migrating VMs.
- Performance Evaluation Technique: What technique is used to evaluate the performance of the proposed system.
- Objective: What are the aspects that the algorithm aims to optimize.

6.6 Conclusion and Future Directions

Although, CDC is the hardware backbone of Cloud-based services, all the undertaken operations in CDC are originated and managed by the Cloud Orchestration software [66], also known as CRMS. CRMS is the key enabler of all types of Cloud centric services rendered towards CSUs. One major concern with CDC is that CDC consumes humongous amount of energy. Energy consumption of CDC is impossible to reduce, if this aspect is not carefully considered while designing the core modules to fulfill the functionalities of CRMS. One of the core functionalities of CRMS is dynamic load balancing. However, the authors of [67] proposed that energy-

Table 6.1 Different aspects of the notable recent VMC algorithms

Research project name	VMC type	VMC decision process	Resource considered	Source PM selection strategy	VM selection criteria	Application of prediction technique	Destination PM selection strategy	Performance evaluation strategy	Objective
Security aware and energy-efficient virtual machine consolidation in cloud computing systems [28]	DVMC	Centralized	CPU	Static and adaptive threshold based	RC, MMT, BPG, MC	Non-predictive	Greedy heuristic	Simulation using CloudSim [49]	Security aware
Dynamic virtual machine consolidation for improving energy-efficiency in cloud data centers [29]	DVMC	Centralized	CPU	Adaptive threshold based	RC, MMT, BPG, MC	Non-predictive	Greedy heuristic	Simulation using CloudSim	Network efficiency aware
Dynamic routing and virtual machine consolidation in green clouds [30]	DVMC	Centralized	CPU	Adaptive threshold based	RC, MMT, HPG, MC	Non-predictive	Greedy heuristic	No performance evaluation	Network efficiency aware
Hierarchical, portfolio theory-based virtual machine consolidation in a compute cloud [24]	DVMC	Centralized	Single resource type which is presented by a numerical value	Adaptive threshold based	RC, MMT, HPG, MC	Non-predictive	Greedy heuristic	Simulation-based performance evaluation	Minimization of intercluster VM migration
Joint VM switch consolidation for energy-efficiency in data centers [45]	DVMC	Centralized	CPU	Threshold-free approach	CVS	Non-predictive	Greedy heuristic	Simulation with real cloud workload traces	Minimization of active PMs and active switch
Thermal aware workload consolidation in cloud data centers [34]	DVMC	Centralized	CPU	Threshold-free approach	RC	Non-predictive	Meta-heuristic	Simulation-based performance evaluation	Minimization of PM and network related energy

Optimizing virtual machine consolidation in virtualized data centers using resource sensitivity [25]	SVMC	Centralized	CPU, memory, and storage			Non-predictive	A mathematical model has been proposed	Simulation in Matlab	SLA violation aware
Virtual machine consolidation with multiple usage prediction for energy-efficient cloud data centers [50]	DVMC	Centralized	CPU, memory, and network bandwidth	Static and adaptive threshold	The VM with highest resource demand	Predictive	Greedy heuristic	Simulation with real cloud workload traces	SLA violation aware
An efficient resource utilization technique for consolidation of virtual machines in cloud computing environments [40]	DVMC	Centralized	CPU	Adaptive threshold	RC, MMT, HPG, MC, MVM	Non-predictive	Greedy heuristic	Simulation in CloudSim	SLA violation aware
Energy and migration cost-aware dynamic virtual machine consolidation in heterogeneous cloud data centers [51]	DVMC	Centralized	CPU, memory, and network bandwidth	Threshold-free approach	RC	Non-predictive	Both greedy heuristics and meta-heuristic algorithms	Simulation with real cloud workload traces	Minimization of active PMs while considering the cost of VM migrations
Cache contention aware virtual machine placement and migration in cloud data centers [26]	SVMC	Centralized	CPU			Predictive	Greedy heuristic	Simulation with real cloud workload traces	Cache contention aware
Improved virtual machine migration approaches in cloud environment [52]	DVMC	Centralized	CPU	Over-utilized and under-utilized PMs are selected as source PMs	MMT	Non-predictive	Greedy heuristics	Simulation in CloudSim with real cloud workload traces	SLA violation aware

(continued)

Table 6.1 (continued)

Research project name	VMC type	VMC decision process	Resource considered	Source PM selection strategy	VM selection criteria	Application of prediction technique	Destination PM selection strategy	Performance evaluation strategy	Objective
Self-adaptive resource management system in IaaS clouds [37]	DVMC	Centralized	CPU, memory	Adaptive threshold	MMT	Non-predictive	Greedy heuristics	Simulation with real cloud workload traces	SLA violation aware
Energy-aware consolidation in cloud data centers using utilization prediction model [46]	DVMC	Centralized	CPU, memory	Static threshold	MMT, VM with minimum resource demand	Predictive	Greedy heuristics	Simulation with real cloud workload traces	SLA violation aware
Robust server consolidation: coping with peak demand underestimation [53]	Both SVMC and DVMC	Centralized	CPU, memory	Threshold-free	All VMs from least power-efficient PMs	Predictive	Greedy heuristic	Simulation with real cloud workload traces	SLA violation aware
Achieving intelligent traffic-aware consolidation of virtual machines in a data center using learning automata [48]	DVMC	Centralized	Network bandwidth	Threshold-free	CVS	Non-predictive	Meta-heuristic	Simulation in CloudSim with real cloud workload traces	Minimize network traffic
Energy optimized VM placement in cloud environment [55]	SVMC	Centralized	CPU			Non-predictive	Greedy heuristics	Simulation in CloudSim	Restrict VM migration and minimize total number of PMs
GLAD: Distributed dynamic workload consolidation through gossip-based learning [32]	DVMC	Centralized	CPU	Threshold-free	VM with least migration cost	Predictive	One of the neighbor PMs is selected	Simulation in PeerSim [54] with real cloud workload traces	Restrict VM migration and minimize total number of PMs

A consolidation strategy supporting resources oversubscription in cloud computing [56]	SVMC	Centralized	CPU			Non-predictive	Greedy heuristics	Simulation in CloudSim	Restrict VM migration and minimize total number of PMs
Bayesian networks-based selection algorithm for virtual machine to be migrated [42]	DVMC	Centralized	CPU	Threshold-free	VM with least migration cost	Predictive	One of the neighbor PMs is selected	Simulation in PeerSim with real cloud workload traces	Restrict VM migration and minimize total number of PMs
Performance-aware server consolidation with adjustable interference levels [23]	DVMC	Centralized	CPU	Static threshold which denotes a level of VM interference level caused by VMC		Non-predictive	Greedy heuristics	Simulation-based performance evaluation	Minimize active PMs while limiting VM interference
A gossip-based dynamic virtual machine consolidation strategy for large-scale cloud data centers [31]	DVMC	Distributed	CPU	Adaptive threshold		Non-predictive	Greedy heuristic	Simulation in PeerSim [54]	SLA violation aware
Energy-efficient migration and consolidation algorithm of virtual machines in data centers for cloud computing [57]	DVMC	Centralized	CPU, storage	Static threshold	Over-loaded PMs: VM with highest resource demand. Under-loaded PMs: All VMs	Non-predictive	Meta-heuristic	Simulation in CloudSim	SLA violation aware
SOC: Satisfaction-oriented virtual machine consolidation in enterprise data centers [58]	SVMC	Centralized	All types of resources required by a VM			Non-predictive	The PM that best matches with IT manager's prefer-ences	Simulation-based performance evaluation	SLA violation aware

(continued)

Table 6.1 (continued)

Research project name	VMC type	VMC decision process	Resource considered	Source PM selection strategy	VM selection criteria	Application of prediction technique	Destination PM selection strategy	Performance evaluation strategy	Objective
Virtual machine placement optimizing to improve network performance in cloud data centers [59]	SVMC	Centralized	Network bandwidth			Non-predictive	Meta-heuristic	Simulation in C++	Minimize total network traffic and maximize link utilization
Virtual machine consolidated placement based on multi-objective biogeography-based optimization [60]	SVMC	Centralized	CPU and memory			Non-predictive	Meta-heuristic	Simulation in C++	Minimize active servers and maximize resource utilization
An energy-aware heuristic framework for virtual machine consolidation in cloud computing [61]	DVMC	Centralized	CPU	Static threshold	RC, MMT, MC, MVM	Non-predictive	Greedy heuristics	Simulation-based performance evaluation	Minimize active PMs and SLA violations while limiting VM migration
Dynamic virtual machine consolidation for energy-efficient cloud data centers [62]	DVMC	Centralized	CPU	Static threshold	VMs from under-utilized PMs	Predictive	Meta-heuristic	Simulation-based performance evaluation	Minimize active servers and maximize resource utilization
Correlation-based virtual machine migration in dynamic cloud environments [63]	DVMC	Centralized	CPU	Static threshold	MC	Non-predictive	Greedy heuristics	Simulation-based performance evaluation	SLA violation aware
VM consolidation approach based on heuristics, fuzzy logic, and migration control [64]	DVMC	Distributed	CPU	Adaptive threshold	Fuzzy VM selection	Non-predictive	Greedy heuristic	Simulation in PeerSim [54]	SLA violation aware
Server consolidation with minimal SLA violations [65]	DVMC	Centralized	CPU	Static threshold	MMT	Non-predictive	Heuristic	Simulation	SLA violation aware

efficiency is rather possible through load unbalancing or workload concentration while switching the idle nodes off, also known as VMC. Since then, researchers have adopted VMC as a strategy to reduce energy consumption in their research work and state-of-the-art VMC algorithms have been developed through following the directions of past researches.

To the best of our knowledge, critical review on VMC algorithms is not available in the existing literature. Hence, in this paper, we have critical reviewed multitude of VMC algorithms with varied viewpoints, as we have also highlighted different aspects of most contemporary VMC algorithms published in 2016 and onwards which are not analyzed in any currently published surveys. Furthermore, we have analyzed notable VMC algorithms published before 2016 which has provided prominent research directions. The summary of our findings is as follows:

- Both SVMC and DVMC algorithms are crucial for energy-efficiency of CDC. SVMC is the first step towards limiting the energy consumption, while DVMC further minimizes the energy consumption while increasing the resource utilization of CDC.
- Centralized DVMC algorithms have been found as more popular than distributed DVMC algorithms. For P2P networks, distributed DVMC algorithms are useful. Although, distributed DVMC algorithms are more robust and reliable in case of hardware failure; however, it poses more network overhead compare to centralized DVMC algorithms.
- In comparison with threshold-free approach, threshold-based approach is more time efficient. However, since threshold-based approach limits the search space by selecting PMs based on the threshold value, therefore stagnation in local minima or local maxima may arise.
- One drawback of VMC is that SLA violation may arise because of aggressive consolidation. Static threshold-based DVMC algorithms cannot control the SLA violation. In contrast, adaptive threshold-based DVMC algorithms limits SLA violation by prior migration VMs from potential overloaded PMs. However, in terms of minimization of energy consumption, adaptive threshold-based DVMC algorithm is less efficient and it causes more number of VM migrations.
- MMT is the most widely used VM selection strategy, as with the decrease of migration time, service disruption time decreases which certainly lowers the SLA violation.
- A wide range of prediction techniques have been proposed in the literature to estimate the future resource demand of VMs as well as future resource utilization of PMs, as predictive DVMC algorithms minimize SLA violation more than non-predictive DVMC algorithms. However, predictive DVMC algorithms exert the overhead of more VM migrations.
- Unlike meta-heuristic algorithms, greedy-based heuristic algorithms may become stagnant in local minima or local maxima, yet these heuristic algorithms provide acceptably sub-optimal solution in quick time. Among all the meta-heuristics, evolutionary algorithms have appeared to be most popular. Considering both heuristics and meta-heuristics, modified version of best fit decreasing is found as most prevalent destination PM selection algorithm.

From our extensive study, we have found the following potential research openings which are yet to be explored:

- Adaptive threshold-based DVMC algorithms and predictive DVMC algorithms minimize SLA violation by prior migration of VMs from those PMs which are identified as potentially overloaded PMs. However, SLA violation is intrinsic in VM migration, because of unavoidable temporary service halt at some point of migration of a VM. Moreover, VM migration increases network traffic and increases network energy consumption. Existing adaptive threshold-based DVMC algorithms and predictive DVMC algorithms do not consider the overhead of VM migration. To address this gap, incorporation of a balanced approach is highly essential which limits VM migration and yet can manage an acceptable SLA violation rate.
- The effectiveness of predictive DVMC algorithms hinges on the actual correlation between past and future resource demand and the quality of adjustment with a specific workload type. However, different VMs co-hosted in a single PM have varied behavioral pattern in terms of resource demand. Thus, no single prediction technique would fit for all PMs, whereas existing predictive DVMC algorithms apply a common prediction techniques for all PMs. Furthermore, there is always a possibility of inaccurate prediction because of potential mismatches between past and present. Hence, there exists a research gap which is yet to be addressed.
- Apart from MMT, rest of the VM selection algorithms only consider CPU demand of VMs and ignore the memory, network bandwidth, and disk I/O requirement of VMs. However, as argued by the authors of [44], selecting a VM only based on CPU will cause saturation in terms of CPU and can lead towards no further improvement in utilization while leaving other types of resource under-utilized. It is highly challenging to determine a single converging point representing the equivalent total resource demand of multitude of resource types, while different types of resources represent different dimensions.

References

1. Kleinrock L. A vision for the internet. ST J Res. 2005;2(1):4–5.
2. Buyya R, Yeo CS, Venugopal S, Broberg J, Brandic I. Cloud computing and emerging IT platforms: vision, hype, and reality for delivering computing as the 5th utility. Future Gener Comput Syst. 2009;25(6):599–616.
3. Kaplan JM, Forrest W, Kindler N. Revolutionizing data center energy efficiency. Technical report, McKinsey & Company; 2008.
4. Beloglazov A. Energy-efficient management of virtual machines in data centers for cloud computing [dissertation]. Melbourne, AU: The University of Melbourne; 2013.
5. Koomey J. Growth in data center electricity use 2005 to 2010. A report by Analytical Press, completed at the request of The New York Times. 2011;9.
6. Gartner Estimates I. Industry accounts for 2 percent of global CO_2 emissions. Press release; 2007.

7. Buyya R, Beloglazov A, Abawajy J. Energy-efficient management of data center resources for cloud computing: a vision, architectural elements, and open challenges; 2010. arXiv preprint arXiv:10060308.
8. Koomey JG. Estimating total power consumption by servers in the US and the world. Feb 2007.
9. Hopkin J. VMware ESX Server [Image on internet]. OStatic; © 2015. Available from: http://ostatic.com/vmware-esx-server/screenshot/1.
10. Mann ZÁ. Allocation of virtual machines in cloud data centers—a survey of problem models and optimization algorithms. ACM Comput Surv (CSUR). 2015;48(1):11.
11. Pietri I, Sakellariou R. Mapping virtual machines onto physical machines in cloud computing: a survey. ACM Comput Surv (CSUR). 2016;49(3):49.
12. Kaur S, Bawa S, editors. A review on energy aware VM placement and consolidation techniques. In: International conference on inventive computation technologies (ICICT). IEEE; 2016.
13. Madhan ES, Srinivasan S, editors. Energy aware data center using dynamic consolidation techniques: a survey. In: Proceedings of IEEE international conference on computer communication and systems ICCCS14. 20–21 Feb 2014.
14. Pires FL, Barán B, editors. A Virtual machine placement taxonomy. In: 2015 15th IEEE/ACM International Symposium on Cluster, Cloud and grid computing. 4–7 May 2015.
15. Ranjana R, Raja J, editors. A survey on power aware virtual machine placement strategies in a cloud data center. In: 2013 international conference on green computing, communication and conservation of energy (ICGCE). IEEE; 2013.
16. Varasteh A, Goudarzi M. Server consolidation techniques in virtualized data centers: a survey. IEEE Syst J. 2015;11(2):772–83.
17. Ahmad RW, Gani A, Hamid SHA, Shiraz M, Yousafzai A, Xia F. A survey on virtual machine migration and server consolidation frameworks for cloud data centers. J Netw Comput Appl. 2015;52:11–25.
18. Choudhary A, Rana S, Matahai KJ. A critical analysis of energy efficient virtual machine placement techniques and its optimization in a cloud computing environment. Procedia Comput Sci. 2016;78:132–8.
19. Masdari M, Nabavi SS, Ahmadi V. An overview of virtual machine placement schemes in cloud computing. J Netw Comput Appl. 2016;66:106–27.
20. Usmani Z, Singh S. A survey of virtual machine placement techniques in a cloud data center. Procedia Comput Sci. 2016;78:491–8.
21. Beloglazov A, Buyya R. Optimal online deterministic algorithms and adaptive heuristics for energy and performance efficient dynamic consolidation of virtual machines in cloud data centers. Concurr Comput. 2012;24(13):1397–420.
22. Ferdaus MH, Murshed M. Energy-aware virtual machine consolidation in IaaS cloud computing. In: Cloud computing. Berlin: Springer; 2014. p. 179–208.
23. Jersak LC, Ferreto T. Performance-aware server consolidation with adjustable interference levels. In: Proceedings of the 31st annual ACM symposium on applied computing. Pisa, Italy. 2851625: ACM; 2016. p. 420–5.
24. Hwang I, Pedram M. Hierarchical, portfolio theory-based virtual machine consolidation in a compute cloud. IEEE Trans Serv Comput. 2016;PP(99):1.
25. Nasim R, Taheri J, Kassler AJ, editors. Optimizing virtual machine consolidation in virtualized datacenters using resource sensitivity. In: 2016 IEEE international conference on cloud computing technology and science (CloudCom). 12–15 Dec 2016.
26. Chen L, Shen H, Platt S, editors. Cache contention aware virtual machine placement and migration in cloud datacenters. In: 2016 IEEE 24th international conference on network protocols (ICNP). IEEE; 2016.
27. Ferdaus MH. Multi-objective virtual machine management in cloud data centers. Melbourne: Monash University; 2016.
28. Ahamed F, Shahrestani S, Javadi B, editors. Security aware and energy-efficient virtual machine consolidation in cloud computing systems. In: 2016 IEEE Trustcom/BigDataSE/ISPA. 23–26 Aug 2016.

29. Deng D, He K, Chen Y, editors. Dynamic virtual machine consolidation for improving energy efficiency in cloud data centers. In: 2016 4th international conference on cloud computing and intelligence systems (CCIS). 17–19 Aug 2016.
30. Fioccola GB, Donadio P, Canonico R, Ventre G, editors. Dynamic routing and virtual machine consolidation in green clouds. In: 2016 IEEE international conference on cloud computing technology and science (CloudCom). 12–15 Dec 2016.
31. Masoumzadeh SS, Hlavacs H. A gossip-based dynamic virtual machine consolidation strategy for large-scale cloud data centers. In: Proceedings of the third international workshop on adaptive resource management and scheduling for cloud computing, Chicago, IL, USA. 2962565: ACM; 2016. p. 28–34.
32. Khelghatdoust M, Gramoli V, Sun D, editors. GLAP: distributed dynamic workload consolidation through gossip-based learning. In: 2016 IEEE international conference on cluster computing (CLUSTER). IEEE; 2016.
33. Beloglazov A, Abawajy J, Buyya R. Energy-aware resource allocation heuristics for efficient management of data centers for cloud computing. Future Gener Comput Syst. 2012;28(5):755–68.
34. Marcel A, Cristian P, Eugen P, Claudia P, Cioara T, Anghel I, et al., editors. Thermal aware workload consolidation in cloud data centers. In: 2016 IEEE 12th international conference on intelligent computer communication and processing (ICCP). 8–10 Sept 2016.
35. Marotta A, Avallone S, editors. A Simulated annealing based approach for power efficient virtual machines consolidation. In: 2015 IEEE 8th international conference on cloud computing. June 27–July 2 2015.
36. Farahnakian F, Ashraf A, Pahikkala T, Liljeberg P, Plosila J, Porres I, et al. Using ant colony system to consolidate vms for green cloud computing. IEEE Trans Serv Comput. 2015;8(2):187–98.
37. Farahnakian F, Bahsoon R, Liljeberg P, Pahikkala T, editors. Self-adaptive resource management system in IaaS clouds. In: 2016 IEEE 9th international conference on cloud computing (CLOUD). June 27–July 2 2016.
38. Farahnakian F, Liljeberg P, Plosila J, editors. LiRCUP: Linear regression based CPU usage prediction algorithm for live migration of virtual machines in data centers. In: 2013 39th euromicro conference on software engineering and advanced applications. IEEE; 2013.
39. Pascual JA, Lorido-Botrán T, Miguel-Alonso J, Lozano JA. Towards a greener cloud infrastructure management using optimized placement policies. J Grid Comput. 2015;13(3):375–89.
40. Selim GEI, El-Rashidy MA, El-Fishawy NA, editors. An efficient resource utilization technique for consolidation of virtual machines in cloud computing environments. In: 2016 33rd national radio science conference (NRSC). 22–25 Feb 2016.
41. Abdi H. Multiple correlation coefficient. Richardson: The University of Texas at Dallas; 2007.
42. Yan C, Li Z, Yu X, Yu N, editors. Bayesian networks-based selection algorithm for virtual machine to be migrated. In: 2016 IEEE international conferences on big data and cloud computing (BDCloud), Social computing and networking (SocialCom), Sustainable computing and communications (SustainCom)(BDCloud-SocialCom-SustainCom). IEEE; 2016.
43. Tanenbaum AS. Distributed operating systems. New Delhi: Pearson Education India; 1995.
44. Ferdaus MH, Murshed M, Calheiros RN, Buyya R, editors. Virtual machine consolidation in cloud data centers using ACO metaheuristic. In: European conference on parallel processing. Springer; 2014.
45. Ma L, Liu H, Leung YW, Chu X, editors. Joint VM-switch consolidation for energy efficiency in data centers. In: 2016 IEEE global communications conference (GLOBECOM). 4–8 Dec 2016.
46. Varasteh A, Goudarzi M. Server consolidation techniques in virtualized data centers: a survey. IEEE Syst J. 2015;11(2):772–83.
47. Dabbagh M, Hamdaoui B, Guizani M, Rayes A. Toward energy-efficient cloud computing: Prediction, consolidation, and overcommitment. IEEE Netw. 2015;29(2):56–61.

48. Jobava A, Yazidi A, Oommen BJ, Begnum K, editors. Achieving intelligent traffic-aware consolidation of virtual machines in a data center using learning automata. In: 2016 8th IFIP international conference on new technologies, mobility and security (NTMS). IEEE; 2016.
49. Calheiros RN, Ranjan R, Beloglazov A, De Rose CA, Buyya R. CloudSim: a toolkit for modeling and simulation of cloud computing environments and evaluation of resource provisioning algorithms. Softw Pract Exp. 2011;41(1):23–50.
50. Nguyen TH, Francesco MD, Yla-Jaaski A. Virtual machine consolidation with multiple usage prediction for energy-efficient cloud data centers. IEEE Trans Serv Comput. 2017;PP(99):1.
51. Wu Q, Ishikawa F, Zhu Q, Xia Y. Energy and migration cost-aware dynamic virtual machine consolidation in heterogeneous cloud datacenters. IEEE Trans Serv Comput. 2016;PP(99):1.
52. Choudhary A, Govil MC, Singh G, Awasthi LK, Pilli ES, Kumar N, editors. Improved virtual machine migration approaches in cloud environment. In: 2016 IEEE international conference on cloud computing in emerging markets (CCEM). 19–21 Oct 2016.
53. Grimes D, Mehta D, O'Sullivan B, Birke R, Chen L, Scherer T, et al., editors. Robust server consolidation: coping with peak demand underestimation. In: 2016 IEEE 24th international Symposium on modeling, analysis and simulation of computer and telecommunication systems (MASCOTS). IEEE; 2016.
54. Montresor A, Jelasity M, editors. PeerSim: a scalable P2P simulator. In: IEEE ninth international conference on Peer-to-Peer computing, 2009 P2P'09. IEEE; 2009.
55. Kaur A, Kalra M, editors. Energy optimized VM placement in cloud environment. In: 2016 6th international conference cloud system and big data engineering (confluence). IEEE; 2016.
56. Liu Y, editor A consolidation strategy supporting resources oversubscription in cloud computing. In: 2016 IEEE 3rd international conference on cyber security and cloud computing (CSCloud). IEEE; 2016.
57. Li H, Zhu G, Cui C, Tang H, Dou Y, He C. Energy-efficient migration and consolidation algorithm of virtual machines in data centers for cloud computing. Computing. 2016;98(3):303–17.
58. Li X, Ventresque A, Murphy J, Thorburn J. SOC: Satisfaction-Oriented Virtual Machine Consolidation in Enterprise Data Centers. Int J Parallel Program. 2016;44(1):130–50.
59. Dong J-k, Wang H-b, Li Y-Y, Cheng S-d. Virtual machine placement optimizing to improve network performance in cloud data centers. J China Univ Posts Telecommun. 2014;21(3): 62–70.
60. Zheng Q, Li R, Li X, Shah N, Zhang J, Tian F, et al. Virtual machine consolidated placement based on multi-objective biogeography-based optimization. Future Gener Comput Syst. 2016;54:95–122.
61. Cao Z, Dong S. An energy-aware heuristic framework for virtual machine consolidation in Cloud computing. J Supercomput. 2014;69(1):429–51.
62. Kang D-K, Alhazemi F, Kim S-H, Youn C-H. Dynamic virtual machine consolidation for energy efficient cloud data centers. In: Zhang Y, Peng L, Youn C-H, editors. Cloud computing: 6th international conference, CloudComp 2015, Daejeon, South Korea, October 28–29, 2015, Revised selected papers. Cham: Springer International Publishing; 2016. p. 70–80.
63. Liu L, Zheng S, Yu H, Anand V, Xu D. Correlation-based virtual machine migration in dynamic cloud environments. Photonic Netw Commun. 2016;31(2):206–16.
64. Monil MAH, Rahman RM. VM consolidation approach based on heuristics, fuzzy logic, and migration control. J Cloud Computing. 2016;5(1):8.
65. Patel CA, Shah JS. Server consolidation with minimal SLA violations. In: Behera HS, Mohapatra DP, editors. Computational intelligence in data mining—volume 2. Proceedings of the international conference on CIDM, 5–6 Dec 2015. New Delhi: Springer India; 2016. p. 455–62.
66. Shackleford D. Virtualization and cloud: Orchestration, automation and security gaps [video on the Internet]. 2DeCipher; 2014. [Available from: https://www.youtube.com/watch?v=mjOwQlr1LIk].
67. Pinheiro E, Bianchini R, Carrera EV, Heath T, editors. Load balancing and unbalancing for power and performance in cluster-based systems. In: Workshop on compilers and operating systems for low power. Barcelona, Spain; 2001.

Chapter 7
Energy Saving in Cloud by Using Enhanced Instance Based Learning (EIBL) for Resource Prediction

Sudha Pelluri and Ramachandram Sirandas

7.1 Introduction

Cloud computing as a technology and business enabler has been the most accepted change in this decade. Cloud computing is being widely adopted by many organizations because of cost-effective solutions offered to users, requirements.

In a cloud computing system the cloud consumer is the ultimate stakeholder. A cloud consumer selects the required service, signs feasible contracts, and uses the service. The cloud consumer is billed for the services requested and makes payments for the services made available to him.

This work contributes to managing IaaS by proposing a new approach for resource prediction and energy saving.

Cloud providers have to play various roles in the process of managing the system. A cloud provider has to provide the requested software/platform/infrastructure services, has to manage technical features required for providing the services, has to provision the services at acceptable SLA levels and protect the security and privacy of the services. Cloud service management is a complex task consisting of among various other activities, resource provisioning.

Provisioning of resources is a challenging issue being faced by the service providers in Cloud, because the requests come from numerous users, requirements dynamically change and there is no specific pattern, trend, or seasonality in the resource usage of users on Cloud. The Google cloud usage data published as trace as described in Sect. 7.2.5 has been studied and the usage of resources and resource requested is shown in Fig. 7.1.

S. Pelluri (✉) • R. Sirandas
Department of Computer Science and Engineering, University College of Engineering, Osmania University, Hyderabad, Telangana, India
e-mail: sudhapv23@gmail.com

W. Rivera (ed.), *Sustainable Cloud and Energy Services*,
DOI 10.1007/978-3-319-62238-5_7

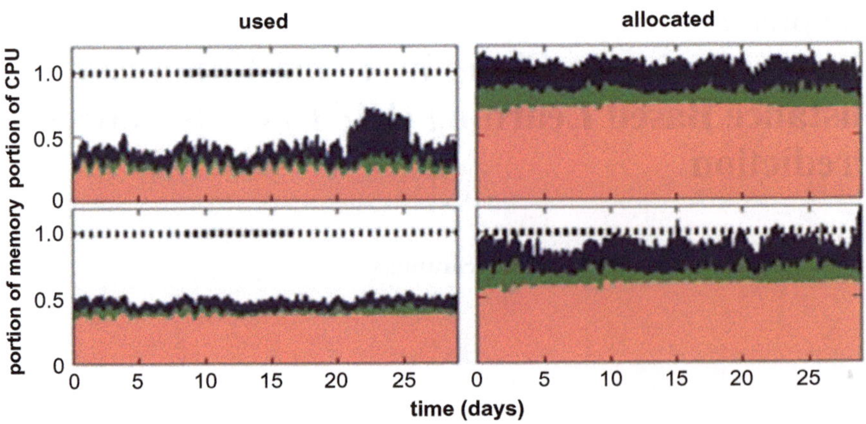

Fig. 7.1 Used resource vs allocated resource of Google Cluster Data *(From [1])*

This shows that resource requirements of the users need to be met **as per their use rather than what users request for**. The real challenge is to be able to provide to the users such quantity of **resources that they will actually use.**

Data Centers are used to house the large compute and storage resources to be made available to users by the cloud providers. Cloud computing applications run over multiple computers connected by a network. How resource management relates to power consumption and power saving is very important for modern day businesses. Google is the first major web company that has revealed its power consumption information. Google uses 260 million watts continuously across the globe as mentioned in their blog—Google Green. According to Koomey, data centers comprised 1.3% of the global energy usage in 2010. Similarly other major cloud providers use huge amount of power. Assuming 3-year server and 15-year other infrastructure cost consolidation, energy-related costs are estimated to amount to 41.6% of operation cost of large-scale data centers as discussed by Hamilton in [2]. Servers are the dominant cost. Power is only 23% of the total, but power distribution and cooling make up 82% of the costs of infrastructure. Cost of building is 12–15%. Hence, overall power consumption costs are considerable.

PUE is ratio of non-computing overhead energy (like cooling and power distribution) to the amount of energy used to power actual machines.Google claims that its data centers have current overhead of just 12%, making their PUE 1.12.

Even with best practices, most companies are not able to reduce some inherent overhead power wastage. Hence, alternatives that help in reducing power to servers need to be identified. This underlines the need for research on resource prediction and energy saving through compute resource units saved.

7.1.1 Motivation

Cloud promises to dynamically provision resources on-demand. Cloud applications are provisioned with specific set of Virtual Machines, comprising of variable units of physical resources (Compute units, memory units, storage units). This necessity to automatically be able to provision resources to users forms the premise for prediction of resource requirements in Cloud. In real cloud usage scenarios, resources requested by users are much higher than what they actually use as specified in [3]. This results in **huge amount of reserved yet unused resources by each user of cloud, for each of his jobs.** This cumulative resource getting wasted results in unnecessary expenditure for customer (he has to pay for each resource unit he reserves for certain period of time) and wasted resource for provider (he could have provisioned resources to other users) for each request. This is a serious problem in cloud resource provisioning.

One of the recent works on scheduling for cloud data centers on Google trace data, (the trace data that has been considered in proposed work), as described in [4], authors have shown that **energy consumed by a machine is proportional to its CPU requests. The more the CPU requests, the larger the frequency scaling factor, which results in cubic increment in power consumption.** Hence, when CPU requests are reduced, by being able to somehow request only that quantum that would be required, it would result in more saving of energy per machine and naturally more the machines, greater the savings.

There is a need to solve real time problem of correctly estimating requirements of resources for each job which would benefit both current cloud providers and cloud consumers.

As another important insight, currently resources are provisioned by the cloud providers (Google compute Engine, Amazon AWS, Rightscale, etc.) **only based on users requests**. Provider tries to optimally use the resources available with them while being able to meet the SLAs (Service Level Agreements) of various users. Currently, mechanisms that try to continuously provision requested resources are said to be **reactive**. Such mechanisms use threshold based rules to detect and react on resource shortages, e.g., time, performance, sum total of minimum available resources. With this resource allocation approach, to overcome even small resource shortages, it takes time of order of few minutes. This is very costly for applications which need frequent scaling. The mean instantiation latency in Amazon EC2 [5] is around 2 min [6]. In experiments conducted as specified in [7], it takes additional 2 min for a Cassandra server [8] to reach its maximum throughput. Thus, reactive elastic resource provisioning scheme like in Amazon EC2 [5], RightScale [9] etc. needs some improvement. **Predictive resource provisioning** can save the setup time that would be required. Predictive approaches for resource requirements enable better utilization of resources, lowering cost for users and overall better management of cloud resources. Predictive resource management approach ensures that all the resource units allocated to different users are efficiently used and saves energy consumed. The loss due to idle power loss can also be saved. The price that providers have to spend comes down due to power saved. Such prediction is useful to meet the objective of GREEN COMPUTING.

7.1.2 Problem Definition

This work proposes a proactive approach for resource prediction that uses resource usage of other users and its own other jobs in the cluster and predicts future requirements without user's intervention. This work intends to show the huge energy savings that can be obtained by cloud providers by using the stated prediction approach.

The major expected benefits of the proposed system are:

1. Efficient capacity planning while ensuring proper scaling and reduced wastage of resources.
2. Reduction of energy consumption. For large companies like Google, a 3% reduction in energy cost can be converted to million dollars in cost saving [10] per year.
3. Reduction of carbon footprint by switching off the unnecessary machines using resource requirement information available beforehand.

However, prediction of host load in cloud is challenging because it fluctuates drastically at small timescales.

7.1.3 Challenges

Resource provisioning and energy estimation gets complicated in the Cloud scenario as compared to existing Web based and Grid based systems because:

1. Resources are requested by the user in real time. The resources are requested when the application starts and this information is not available beforehand. Users **expect Instantaneous Resource availability**. The resources are to be made available to the users as the application execution proceeds. This is difficult to implement by the Cloud Provider because of erratic requests of various users as shown by the Cloud usage trace—Google Trace data as described in [11].
2. To make available resources as per changing requests of users, to enable **dynamic scalability of resources**, reactive approaches are easier to implement but take unacceptable time to provision resources. Predictive approaches serve the purpose of resource reservation but are difficult to implement as they depend on historical data and on cloud, new users without historical information are more common, making proactive provisioning difficult for cloud environment.
3. The trace of Cloud usage shows that it does not lend itself to existing prediction approaches like time series, queuing models Bayesian model, SVM, neural networks, etc. because of absence of patterns, trends, and seasonality. The Auto Correlation Function (ACF), from the Correlogram on Cloud resource usage of Google Cluster data (shown in Fig. 7.2), shows that no specific pettern can be derived (From [12]).

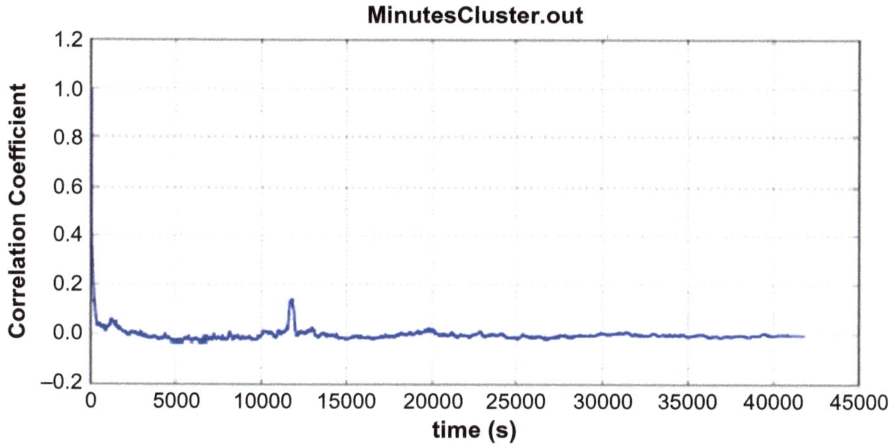

Fig. 7.2 Correlogram of Google Cluster Data

4. The machines that are made available to users by various providers like Google Cloud Platform—Google Compute Engine—available at http://cloud.google.com/compute/docs are quite variable. Google has chosen 2.75 GCEUs to represent the minimum computational capacity of one virtual CPU (a hardware hyper-thread) on Sandy Bridge, Ivy Bridge, or Haswell platforms. **These are different in their energy consumption—idle, full load, and average. Therefore it becomes difficult to exactly estimate the power consumed per user request for CPU**. Recent work by Ian Cutress at Anandtech helps in knowing this power consumption.
5. Energy consumed per job, per task cannot be estimated as an absolute number because of the data obfuscation in Google Cluster data. Dynamic power consumption here based on the resource units consumed gives us energy consumed and energy saved on two models—Intel Xeon 2687 and Intel Xeon 2697 is a relative measure. Similar approach can be employed on various other machines where proportional increase in power consumption by use of compute units is known.

7.1.4 Chapter Organization

This chapter consists of seven sections. In this section, an introduction that outlines the background briefly, brings out the motivation, specifically points out the objectives and challenges of this work has been covered. The main contributions are also enumerated for greater clarity.

In Sect. 7.2, existing work on all related concepts, techniques, approaches, and workload is discussed. This includes only most relevant works though extensive literature survey was actually undertaken to get thorough understanding of important and recent works in all related areas. Section 7.3 includes process and technique of proposed work. Section 7.4 shows the quantified specific results and the analysis of this work. Section 7.5 is significant as it discusses the importance, available solutions, proposed solution, and the results for energy saving. Contributions of this work are explicitly mentioned in Sect. 7.6. Conclusion and future work are included briefly in Sect. 7.7. Finally important references are included.

7.2 Related Work

There has been considerable research on various aspects of Cloud Computing in the last few years and the body of research keeps growing by the day. The entire study of existing systems is exhaustive. Most relevant works have been included in this section.

7.2.1 Resource Provisioning in Cloud

There is considerable body of work that focuses on resource provisioning and resource management while specifically addressing energy requirements. In [13] the authors propose resource allocation policies for the management of multi-tier virtualized cloud systems with an aim to maximizing the profits associated with multiple-class SLAs. Antony Beloglazov et al. [14] propose approaches for QoS at reduced cost. In [15], effective use of middleware architecture to handle various quality of service parameters is discussed. In [16] the authors propose budget and deadline constraints aware approach to provide resources based on requests that users make. They suggest approach for user to estimate his requirements. In [17], the authors suggest techniques for dynamic autonomous resource management in computing clouds. In [18] the authors propose a scheduling algorithm for real time services. In [19, 20] and [21] the authors propose an autonomic resource manager to control the virtualized environment which decouples the provisioning of resources from the dynamic placement of virtual machines. In [20] the authors propose an OVMP algorithm. The algorithm can minimize the cost for hosting VMs on multiple cloud provider environment under demand and cost variations.

7.2.2 Auto-Scaling

Dynamic resource allocation strategy at the VM level is implemented as **horizontal scalability or vertical scalability** in [22]. Vertical Resizing adjusts logical partition of resources (e.g., CPU, Memory, Bandwidth, I/o etc.) in a VM. Amazon AWS

EC2CS and some third party cloud management services like Rightscale [9], AzureWatch [23], Scalr [24], etc. **offer schedule based and rule based auto-scaling mechanisms**. Rule based mechanisms work on user defined triggers by specifying instance scaling limits and corresponding actions. RightScale [9] and AzureWatch [23] use some middleware metrics like database connections, web server requests, name resolution queries, and queue sizes. These scaling indicators do not support deadlines or consider user cost, though they make scaling easier for web based applications. Auto-scaling is Reactive or Predictive.

1. Reactive Auto-Scaling. Cloud Auto-scaling based on user specified triggers, and performance parameters is discussed in [25]. Deadlines and cost are major constraints when using auto-scaling. When Workloads and deadlines keep varying, tasks on the VMs use "Earliest Deadline First" approach. But this approach can be implemented only when scaling is allowed to be reactive because the approach is time consuming.
2. Predictive Auto-Scaling. Capacity planning is complicated in real cloud scenarios because, when planned for average load, there is less cost in initial setup but performance drops when peak load occurs. When capacity is planned for peak workload, resources will not be fully utilized. Present day VM provisioning technology takes few minutes to provision a VM. Distributed resource scaling involves more bundling and provisioning time. For example, the mean instantiate latency in Amazon EC2 is around 2 min [6], takes a while for new server instances to boot up. It takes another 2 min for Cassandra server [8] to reach its maximum throughput. All this adds to the delay. The cloud service provider could prepare all the VMs ahead only if it knew information on future requirements of the users. Basic approaches of workload prediction for auto-scaling is introduced in [26]. The ARMA method is suggested as prediction technique.
   Even though elastic resource provisioning is provided by Infrastructure as a service (Iaas) in clouds by providers like Amazon [5], deciding when to get more resources, how many to get, when application workloads and service level objectives are dynamically changing is quite difficult.

7.2.3 Resource Prediction Approaches

The main work on predictive scaling involves finding the predicted requirement of resources for the users on cloud. A comprehensive text on time series [27] introduces the concepts of time series in detail. Stationary, non-stationary, seasonal, and multivariate time series are explained. An excellent text which gives basic information is [28]. The linear models for prediction of host loads as originally suggested by Box Jenkins by using AR, MA, ARMA, ARIMA, ARFIMA are discussed in [29]. This approach is suitable only when there is a linear relationship between the successive loads, which as we know is a rare possibility in Cloud. Capacity planning is described in [30]. A Sliding window approach of identifying the patterns and trends in demand is used. In [31] the authors have proposed a resource prediction approach by using Double Exponential Smoothing which considers current state and history of resource used. They show how the current state can be modeled in terms of previous state.

Cloud workload can be modeled as a G/G/N queuing model as in [11]. This is used in building a proactive, reactive hybrid controller that uses history and a quick reaction scheme for scaling. Bayesian forecasting techniques with case studies are described in [32]. In [33] the authors have applied the prior probabilities distribution for target states. Training period is taken as first 25 days of the Google Cluster data trace. The remaining 3 days data is taken as test data. The limitation is that all results are subject to accuracy of *prior and joint probability.*

In [34], three different approaches for prediction—Artificial Neural Networks, ARIMA time series, and Regression are used for prediction of spring flow. An integrated approach for predictive elastic scaling which involves multiple approaches combined is discussed in [35]. A combination approach for prediction system is presented in [36]. Multiple type VM demands based on request history, with specific ensembles for each type are used. Aggregated time series based on features like VM type and time of request start/request end are used. The use of input features in prediction is described in [37]. The e-science workflows are computationally expensive.

7.2.4 Energy Conservation

There is focus on different aspects of energy related to cloud like, energy of components of data center, energy-efficient virtual machine allocation, virtual machine migration, task consolidation, energy-efficient task scheduling, and virtual machine power metering and provisioning. Power consumption for servers, storage devices, and network equipment in a data center is discussed in [38]. In [39] the authors present an online provisioning approach for HPC applications. In [40] the authors created a mathematical model for power management for a cloud environment where users have interactive applications such as web services. In [41] a new framework that provides efficient green scalable cloud computing architecture is proposed. Green Data center is a concept being strongly supported by Google. In an attempt to reduce the power consumption at their data centers they have adopted and are advocating a set of best practices. These are available at their site http://google.co.in/green/efficiency/datacenters. One of the recent works on using learning for Data center optimization has been very well received as shown in [42]. Energy stored in multiple Energy Storage Devices (ESD) in Data centers for Demand Response (DR). Currently high density data centers are being built with distributed ESDs, e.g., server-level UPS units in Google, rack and power distribution units in Facebook and Microsoft. These are discussed in [43]. Power measurements for different types of CPU makes and different number of cores is described in [44]. They also have shown that power consumption is rather constant when a process is writing to memory. **In another work [45], fine grained energy consumption, and performance data, different resource allocation strategies, system configurations and workloads are shown. Experimentally they have shown that power consumption by memory units does not change much in workload. This is the main reason why we do not need to focus on energy saved with memory units saved.**

Google tries to save energy during backup by reducing losses due to conversion between central back up unit and servers. Facebook has launched the Open Compute Facebook Project to enable data centers to be 38% more efficient and 24% less expensive to build.Yahoo has a new facility called "Computing Coop." This structure overcomes the need for chillers. This is expected to create a PUE (Power Usage Effectiveness) of 1.1. There are efforts in this direction from various quarters. Our approach to energy conservation is described in Sect. 7.5.3.

7.2.5 Google Cluster Data

Real cloud usage data has been released by Google in 2011. We have used the Google Trace Ver 2.0, which is a 29-day trace on 12k-machine cell in May 2011.

The work by Mishra et al. [46] captures the heterogeneity and dynamicity of data in Google trace. In [1] by Charles Reiss, consolidated cloud environments are constructed from numerous machine classes. In [47], the authors compare the two workloads—GridMix3 and Yahoo production cluster by using k means clustering approach. By using time series the data of real cloud usage trace from IBM hosted cloud is studied for both CPU and Memory usage is [48]. Synthetic workload was generated in [49] by using the resource utilization and task wait time. An important paper that discusses about scheduling and dependencies is [50] by Sharma. A useful paper that helps in understanding the workload (for jobs, tasks) and host load (at machine level) in a Google Data Center in comparison to the Grid system is provided in [51] by Sheng Di. In [52] the authors have provided a reusable approach for characterizing cloud workload obtained from the Google Compute trace of 29 days. By using an approach called autonomous cooperative cloud based platforms (ACCP), in [53], the authors have shown the advantage of cooperative approach in Inter-Cloud environments. A simulation based evaluation based on SCIENCE CLOUD under a model of realistic Google Cluster data set [54] has been used. They have tried to show that this approach gives better performance as compared to selfish approach in large clusters.

7.2.6 Issues with Existing Prediction Techniques

Predictive scaling is shown to be an encouraging alternative to reactive scaling specifically when we expect resources to be available beforehand and do not want the users to wait. Based on the current research body of work available the following conclusions are reached:

1. Most of the techniques used today in resource prediction are based on prediction for Web workloads. **The main difference between Web workloads and Cloud Workloads is the absence of trend, seasonality, and diurnality in Cloud workloads**. Hence, the approaches used in Cloud resource prediction need to be different from those for Web workloads.
2. Current approaches to predict resource requirements are based on availability and processing of large amount of history data. This is not necessarily possible when the prediction is to be done for initial start cases in any system.
3. Fetching and Processing of history data requires time even with best algorithm and data access techniques. **This time lag is difficult to afford for real time prediction in Cloud resource provisioning.**
4. Some of the combined approaches shown by other researches need to select specific approaches based on available data. But this would also show some lag time and not enable prediction in real time.
5. The work on energy saving has mostly been coarse grained, whereas our approach is more fine grained. We inspect the possible energy saving that can be obtained by saving resource units, more specifically CPU units for each job of individual user.

7.3 Solution and Implementation

The need to find efficient predictive scaling approaches, need for resource and energy saving approaches have led to proposing new technique—Enhanced Instance Based Learning(EIBL). The implementation results obtained are shown to be useful and fairly accurate as described in Sect. 7.4. The basic idea is to be able to predict CPU Usage and Memory usage for one specific user using two variants—Distance Weighted Averaging and Locally weighted Regression. The adoption of these helps in selecting the most profitable variant to implement in real time. The description of the workload that has been used is included to make the understanding of process clear.

7.3.1 Workload Description

Workload that is representative of real Cloud workload has been identified. A real trace of Cloud usage as presented in Google Cluster-usage traces published in Nov. 2011 is used as workload for analyzing the efficiency of our approach [55]. This approach of using real Cloud trace is far more useful than using representative synthetic workloads as done in competing approaches described by other researchers. The workload trace is a 29-day trace with various entries pertaining to real cloud usage. It gives information on 12,583 servers, 25 million tasks grouped into 650,000 jobs, and 925 users. A Google cluster is a set of machines, packed into racks, and connected by a high-bandwidth cluster network. Work arrives at a cell (a set of machines) in the form of jobs. A job is comprised of one or more tasks, each characterized by a set of resource requirements used for scheduling (packing) the tasks onto machines. Each task represents a Linux program, possibly consisting of multiple processes, to be run on a single machine. Tasks and jobs are scheduled onto machines according to the life cycle. Resource requirements and usage data for tasks are derived from information provided by the cell's management system and the individual machines in the cell. A single usage trace typically describes several days of the workload on one of these compute cells. The various tables are indexed by primary keys. The tables published in this trace are—Machine Events Table, Machine Attributes table, Job Events Table, Task Events Table, Task Constraints Table, and Task Resource Usage Table.

From the various tables, the two tables of interest to us are the **Task Events table and Task resource Usage Table**. These tables are merged on ***jobid***. The entries from these tables show the resources requested and the resources used by user.

The reasons for selection of this trace for testing the efficiency of proposed prediction approach are:

1. Trace used in similar works is Real VM trace log of IBM Smart Cloud Enterprise (SCE) product. In SCE trace data, each VM request record contains 21 features such as Customer ID, VM Type, Request Start Time, and Request End Time, etc. However, this trace is actually available only within IBM and described in [56]
2. Some selected universities were given access to Yahoo trace logs. To gain insight on MapReduce workloads 10 months of trace data from the M45 supercomputing cluster, a production Hadoop environment of Yahoo was used. The M45 cluster has approximately 400 nodes, 4000 processors, 3 terabytes of memory, and 1.5 petabytes of disk space. The cluster runs Hadoop, and uses Hadoop on Demand (HOD) to provision virtual Hadoop clusters over the large physical cluster as mentioned in [57]
3. Other traces from Grids like AuverGrid, Nordu Grid, and Web access logs like worldCup 1998 trace are not exact representatives of real cloud usage. Hence those traces are not used in experimentation and testing.

7.3.2 Process Description

The two files—Task events table and Task resource usage table are merged on jobid. IBM SPSS statistics tool is loaded with these tables. The proposed algorithms, Distance Weighted Averaging and Locally Weighted Regression are tested with the merged file using SPSS Statistics. The expected results for resource savings (CPU and Memory) and energy saving are obtained and analyzed. For computing Energy Savings information regarding power consumed by Intel Xeon E5 2687 and Intel Xeon E5 2697 was used.

7.3.3 Enhanced Instance Based Learning

In implementation of Enhanced instance based learning, from the stated trace data, we have used user id, CPU requested, CPU used, memory requested, memory used for various cases. Resource usage prediction for the user, for whom only his amount of **resource requested** information is available is computed. When Workloads and deadlines keep varying, tasks on the VMs use "Earliest Deadline First" approach. But this approach can be implemented only when scaling is allowed to be reactive because the approach is time consuming.

The energy saved by implementing each of the different approaches for resource prediction have been included in proposed Enhanced Instance Based Learning (EIBL) approach. The various steps in the proposed approach are as follows:

1. Resource requirement prediction by:
   (a) Distance Weighted Averaging using Gaussian Kernel
   (b) Locally Weighted Regression
2. Error estimation
   (a) By Measuring Residuals.
   (b) By using paired sample 't' tests

The basic approach is to predict the amount of resource usage for user by using Distance Weighted Averaging (DWA) and Locally Weighted Regression (LWR). The intention is to find the nearest neighbors of specific case (for which resource usage prediction is intended) based on CPU Usage. K Nearest Neighbor (KNN) is needed to do this similarity check. Initially the available information is too huge to start trying to find nearest neighbors whose resource usage values can be useful for prediction for the target case. As mentioned in [58], traditional KNN classification has high calculation complexity to find out similarities between training samples. To overcome this situation in present solution, there is a need to reduce the scope of prediction by localizing the scope to most plausible cases. Mechanism to ensure lesser number of comparisons for identifying most similar resource usage values for the case under consideration is needed. The K-means algorithm is used to address this requirement.

The contribution of this work lies in very specifically identifying how Instance Based Learning approaches can be enhanced to be effectively used on complex Cloud data for resource prediction. More significant is the prediction of the energy savings that can be obtained when the two flavors of Enhanced Instance Based Learning are applied to the actual Cloud Usage Trace.

7.4 Evaluation and Analysis

In this section quantified results for resource and energy savings obtained are included. After having followed the various steps in EIBL, for various users and in different clusters resources are predicted and performance of the approach is measured using **metrics** mentioned below:

1. Resource savings obtained
2. Prediction accuracy
3. Data set size used

These are discussed in detail to show the efficiency of the EIBL approach.

1. **Because of resource prediction by EIBL**
   Saving = Resource Requested – Resource predicted
   CPU Units saved by EBIL = CPU units requested – CPU Predicted
   Memory units saved by EBIL = Memory requested – Memory units Predicted

2. **Without any prediction approach initial ad-hoc method of guessing the need for resource by user results in residual which is computed as:**
   CPU Residual = CPU Requested by user – CPU actually used
   Memory Residual = Memory Requested by user – Memory actually used

3. **After the EIBL prediction approaches are used, we measure accuracy as**
   Residual = Resource Predicted – Resource actually used
   CPU residual = CPU Predicted – CPU used
   Memory residual = Memory Predicted – Memory used

4. **Data Set size used**
   Data set size = size of cluster formed by using K-means on Specific CPU usage, (as obtained from CPU request information). **The proposed solution uses only localized information and does not require large amount of history for training data.**

Various types of Graphs have been used to show the results for different use cases- specifically for users with user id- 148, user id- 169, user id - 412. In some graphs, combined information from the three users have been shown.

7.4.1 *Resource Savings*

One of the goals of our work is to find a prediction mechanism that can reduce the unnecessary excess booking of resource by the user. With the prediction of resource requirement by using our EIBL approach we can show that there is a huge reduction in wastage of resources as compared to ad-hoc method of resource reservation by the user. For users who have multiple instances (many jobs for which resources were requested) in each cluster, resource requirement prediction by both DWA (Distance Weighted Averaging) and LWR (Locally Weighted Regression) is performed.

7.4.1.1 CPU Savings

Initially some graphs to show savings obtained are included to show the importance of DWA and LWR approaches. The five Clusters into which the data file was divided based on k means clustering contain information regarding various users. Cluster 1 and Cluster 3 had only 3 users each, Cluster 4 had 34 users, cluster 2 had 4611 users, and Cluster 5 had 352 users. Hence, only Cluster 2 and Cluster 5 information is used in experiments. Of the various users, only 3 users had significant presence in these two clusters, Cluster 2 and Cluster 5, User id- 148, user id.- 169, and User id- 412 show significant number of cases. So, most of the results being discussed here are for these 3 users only.

The CPU savings for all 3 users with uid 148, uid 169, uid 412 together in same graph are shown.

Sum total of the saving that can be obtained for all the jobs within that cluster for each specific user is computed and shown in Fig. 7.3. **This actually depicts the saving for the user by using DWA and LWR method of prediction.**

7.4.1.2 Memory Savings

Memory savings are shown with the bar graphs as below. One observation that can be made is that quantum of memory saved is not as huge as quantum of CPU saved by the prediction approaches. The results for all 3 users, users with uid 148, uid 169, uid 412 in same graphs are shown.

Individual saving of memory units by both DWA and LWR per job are available. But, these are variable. Sum total of the saving that can be obtained within that cluster for specific user is computed and shown in Fig. 7.4. **This actually depicts the saving for the user by using DWA and LWR method of prediction.**

7.4.1.3 Analysis of Resource Savings Obtained

The results of resource savings obtained are available in comprehensive comparative tables as below.

1. Resource Savings obtained
   (a) Savings for CPU

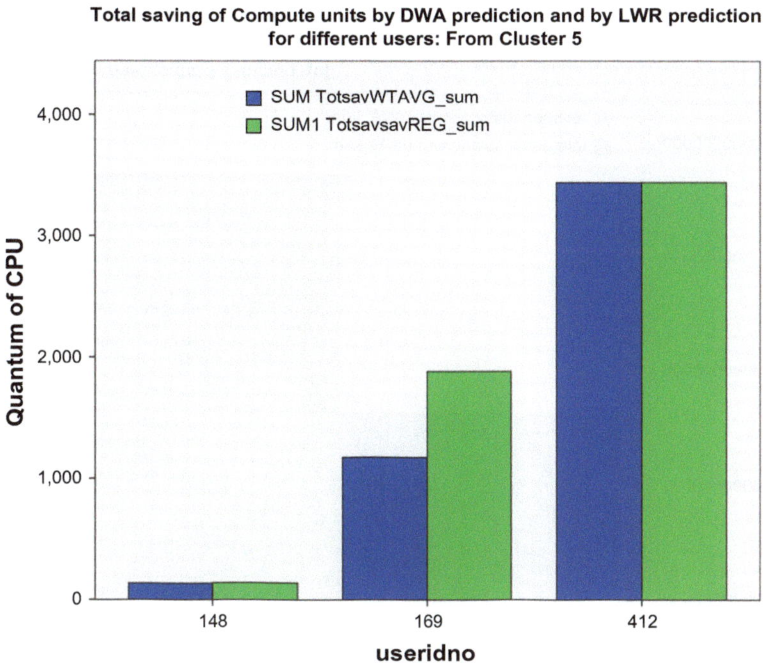

Fig. 7.3 Sum of compute units saved by each job by using DWA and LWR for all 3 users

- Saving in CPU by prediction using DWA.
  The savings and resource wastage reduction is available in a table for all the 3 users in Fig. 7.5.
- Saving in CPU by prediction using LWR
  The savings and resource wastage reduction is available in a table for all the 3 users in Fig. 7.6.

(b) Saving for memory

- Saving in memory by prediction using DWA. The savings and resource wastage reduction is available in a table for all the 3 users in Fig. 7.7.
- Saving in memory by prediction using LWR. The savings and resource wastage reduction is available in a table for all the 3 users in Fig. 7.8.

7.4.2 Accuracy

In order to evaluate the accuracy of prediction, the predicted values are compared against the real values of the resource actually used by the user. This is the advantage of using Google Trace data.

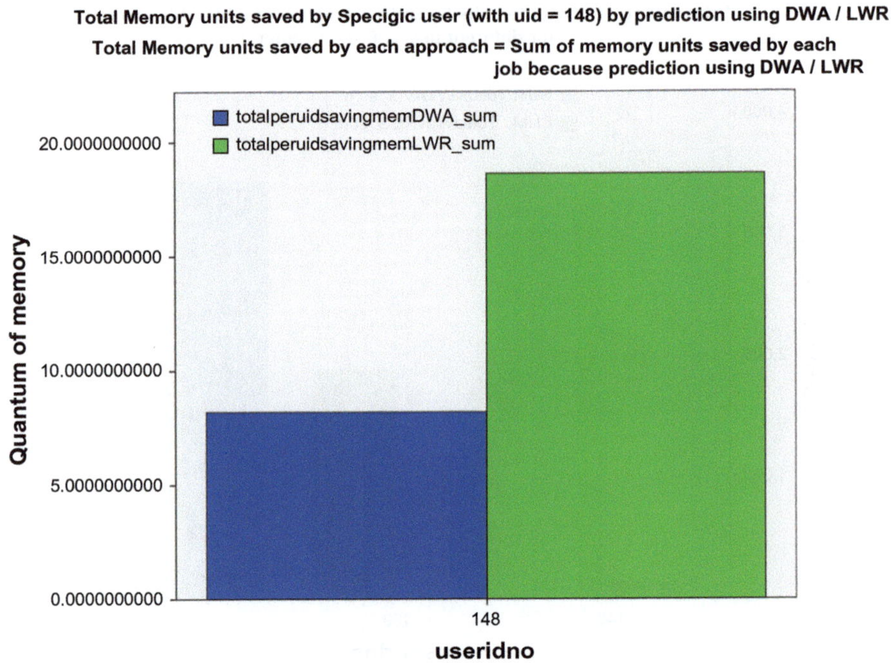

Fig. 7.4 Sum of memory units saved by each job by using DWA and LWR

Based on the readings obtained from the experiments, we can check how near is our prediction to the actual resource used.

Residual = Resource units predicted by our approach – Actual Resource units used by the user

7.4.2.1 Accuracy of CPU Prediction

Residual in CPU Prediction

The information regarding accuracy of Prediction of resource requirements for one user with uid 148 can be shown and then graphs showing accuracy of prediction for various users with uid 148, uid 169, and uid 412 can be shown in same graph.

In measuring estimation error for compute resources, it varies for each job of the user. Hence, mean absolute percentage error in estimation by Ad-hoc, by DWA, by LWR for this user (uid 148) is shown in Fig. 7.9.

	Observed variation as compared with Actual CPU Usage		Percentage reduction in variation	Saving of Compute units
	For Initial Adhoc estimate for resources. (CPU request – CPU Usage) (A1)	For Prediction by DWA (CPU predicted by DWA – CPU usage) (A2)	((A1 – A2) / A1) * 100	Resource units initially requested – Predicted resource by DW
Uid = 148	3.8562	0.24979	93.5224	4.1060
Uid =169	14.980	2.7253	81.8071	12.2526
Uid= 412	370.3162	12.1423	96.7212	382.4585

Fig. 7.5 Sum of compute units saved per job by using DWA

	Observed variation as compared with Actual CPU Usage		Percentage reduction in variation	Saving of Compute units
	For Initial Adhoc estimate for resources. (CPU request – CPU Usage) (A1)	For Prediction by LWR (CPU predicted by LWR – CPU usage) (A2)	((A1 – A2) / A1) * 100	Resource units initially requested – Predicted resource by LWR
Uid = 148	3.8562	0.41088	89.345	4.1088
Uid =169	14.980	9.8421	34.2984	24.8200
Uid= 412	370.3162	12.3400	96.6678	382.6592

Fig. 7.6 Sum of compute units saved per job by using LWR

	Observed variation as compared with Actual Memory Usage		Percentage reduction in variation	Saving of Memory units
	For Initial Adhoc estimate for resources. (Memory request – Memory Usage) (A1)	For Prediction by DWA (Memory predicted by DWA – Memory usage) (A2)	((A1 – A2) / A1) * 100	Resource units initially requested – Predicted resource by DWA
Uid = 148	0.6280404	0.37986268	39.5163	0.248177
Uid =169	13.4812	1.24221	90.7856	12.23894
Uid= 412	43.650755	4.5734	89.523	39.0773

Fig. 7.7 Sum of memory units saved per job by using DWA

	Observed variation as compared with Actual Memory Usage		Percentage reduction in variation	Saving of Memory units
	For Initial Adhoc estimate for resources. (Memory request – Memory Usage) (A1)	For Prediction by LWR (Memory predicted by LWR – Memory usage) (A2)	((A1 – A2) / A1) * 100	Resource units initially requested – Predicted resource by LWR
Uid = 148	0.6280404	0.0635051	89.8876	0.5645353
Uid =169	13.4812	3.1633	76.5355	10.317825
Uid= 412	43.650755	0.480657	98.8988	43.170098

Fig. 7.8 Sum of memory units saved per job by using LWR

7.4.2.2 Accuracy of Memory Prediction

Residual in Memory Prediction

In measuring estimation error for memory resources, it varies for each job of the user. Hence, mean absolute percentage error in estimation by Ad-hoc, by DWA, by LWR for this user (uid 148) is shown in Fig. 7.10

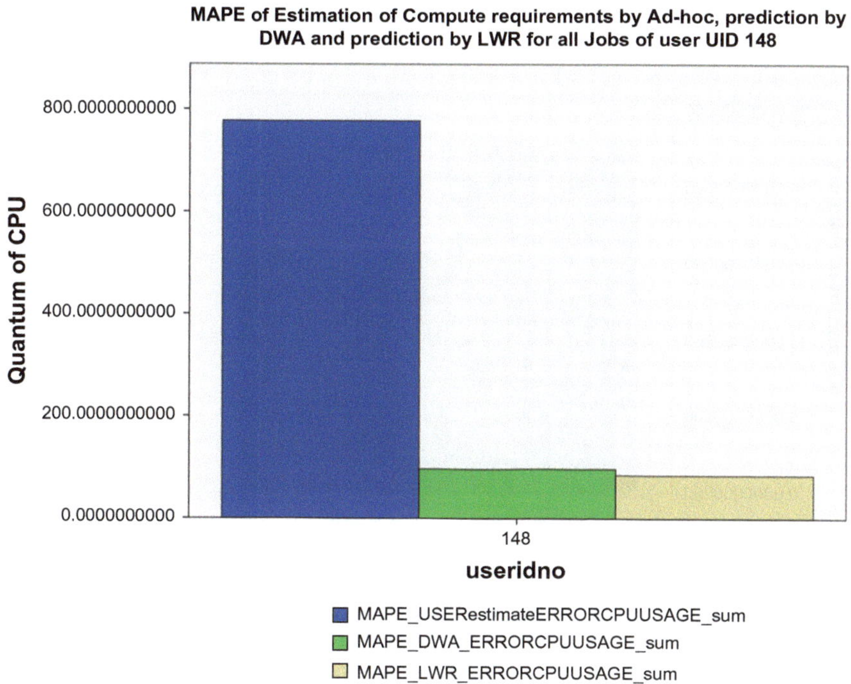

Fig. 7.9 Mean absolute percentage error of CPU estimation by Ad-hoc Resource requests, resource prediction by DWA, resource prediction by LWR

Actual values of residuals can be shown for one cluster, for one user.

Results show the difference by using various approaches in estimating the resource that will be used. This is important to see the difference in estimated error is very much reduced by using DWA and LWR approaches. The line graphs show the residuals for CPU usage by initial users estimate by ad-hoc method, values predicted by using DWA and values predicted by using LWR as seen in Fig. 7.11.

The residuals for users estimate of Memory usage is higher than the residual because of using DWA or LWR as can be seen in Fig. 7.12. This shows that predicted values of resource are much nearer to the actual resource usage values as recorded in Google cluster data tables.

7.4.2.3 Analysis of Accuracy of Prediction

The correction of prediction approaches are measured by the estimation error. Accuracy of prediction measured by Residuals

1. Residual measure for CPU prediction

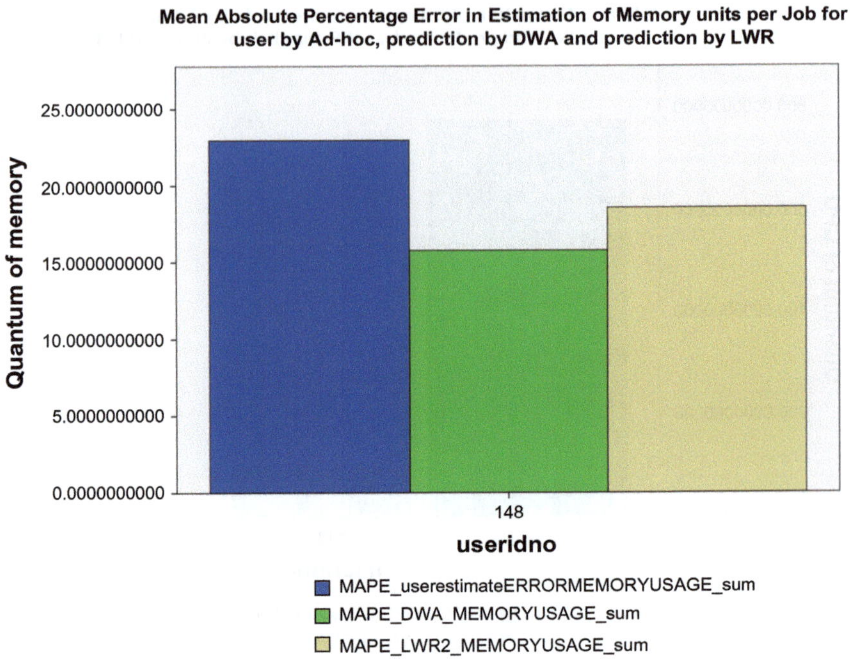

Fig. 7.10 Mean absolute percentage error of Memory estimation by Ad-hoc resource requests, resource prediction by DWA, resource prediction by LWR

(a) Residual measure for CPU prediction using DWA.
Mean Absolute Percentage Error and reduction in error for CPU estimate per job by using DWA is available in a table as shown in Fig. 7.13

(b) Residual measure for CPU prediction using LWR.
Mean Absolute Percentage Error and reduction in error for CPU estimate per job by using LWR is available in a table as shown in Fig. 7.14 and can be easily understood.

2. Residual measure for Memory prediction

(a) Residual measure for Memory prediction using DWA
Mean Absolute Percentage Error and reduction in error for Memory estimate per job by using DWA is available in a table as shown in Fig. 7.15

(b) Residual measure for Memory prediction using LWR.
Mean Absolute Percentage Error and reduction in error for Memory estimate per job by using LWR is available in a table as shown in Fig. 7.16.

[DataSet1] C:\Users\admin\Desktop\New folder\Individualclusterall data\New Folder\ERRRORall.sav

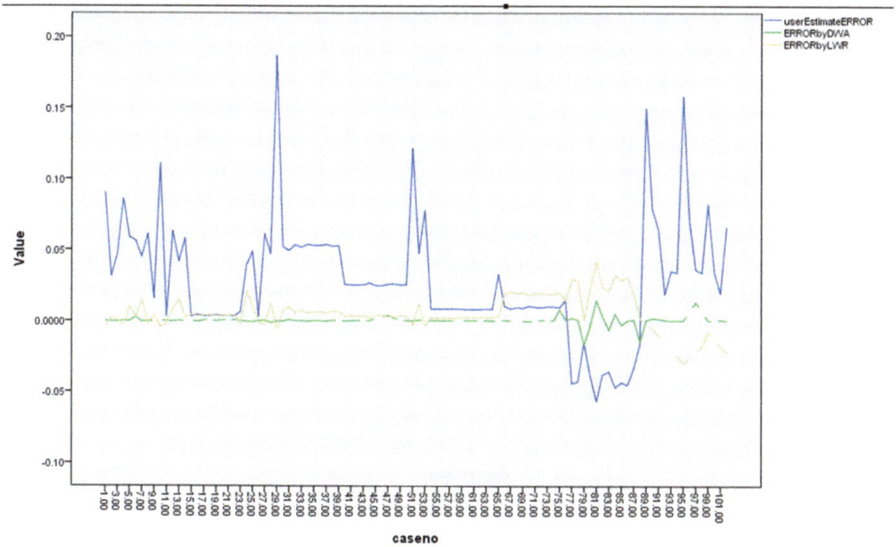

Fig. 7.11 Residuals for CPU usage of users initial estimate, DWA and LWR on a graph

7.4.2.4 Measure of Accuracy Using Paired Sample ‘t’ Test

The paired sample ‘t’ test forms Figs. 7.17 and 7.18 shows that mean difference between the predicted values using DWA from actual CPU usage values and actual memory usage values is not statistically different from 0 (p value observed > 0.05). Hence the results are very near to actual resource usage values when DWA is used.

7.5 Energy Conservation

7.5.1 Importance

The rapid move to Internet-enable devices, use of Software-as-a-Service, growth of Big Data applications and organizations moving their operations to Cloud-Based Systems has accelerated the growth of large-scale Data Centers. The Energy used in a large-scale data center is a matter of great concern for all organizations and nations which are concerned about climate change. Some of the largest and most complex data centers are owned by well-known internet giants like Facebook, Google, and Amazon. Organizations strive to reduce operations and energy costs by implementing a variety of solutions such as relying on renewable energy sources and power saving in data centers.

```
[DataSet2] C:\Users\admin\Desktop\workingcl2\breakuponid\id412.sav
```

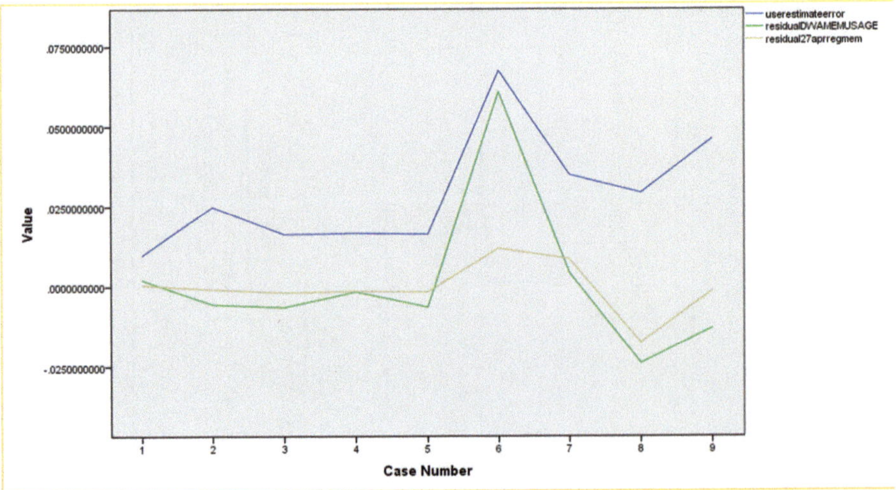


```
DATASET ACTIVATE DataSet3.
COMPUTE Userestimateerror=memoryrequest_mean  -  assignedmemoryusage_mean.
EXECUTE.
GRAPH
  /LINE(MULTIPLE)=VALUE(Userestimateerror residualMEMUSAGE residualREGR27aprilMEMUSE).
```

Fig. 7.12 Residuals for Memory usage of users initial estimate, DWA and LWR on a graph

Caseid	For CPU		
	MAPE in User CPU Estimate	MAPE in prediction by DWA	% reduction in error
For Uid= 148	2357.85	291.81	87.6239
For Uid= 169	5341.19	4.694	99.91
For Uid= 412	2976.85	63.615	97.83

Fig. 7.13 Mean absolute percentage error and reduction in error for CPU estimate per job by using DWA

To be able to provide such services to the users in real time, there are huge **operational challenges in managing a Data Center.** Energy-efficiency is a very important consideration world over. Large organizations have focused on identifying ways for fostering energy-efficient protocols, architectures, and techniques. To

Caseid	For CPU		
	MAPE in User CPU Estimate	MAPE in Prediction by LWR	% reduction in error
For Uid= 148	2357.85	255.35	89.17
For Uid= 169	5341.19	307.21	94.25
For Uid= 412	2976.85	173.44	94.17

Fig. 7.14 Mean absolute percentage error and reduction in error for CPU estimate per job by using LWR

Case id	For Memory		
	Error in User Memory estimate	Error in prediction by DWA	% reduction in error
For Uid= 148	69.5218	47.77	37.71
For Uid= 169	173.6548	57.3372	100.71
For Uid= 412	209.9911	77.7577	62.97

Fig. 7.15 Mean absolute percentage error and reduction in error for Memory estimate per job by using DWA

ensure maximum availability of services to the clients, most of the data centers try to provision more resources so that customers get the resources 24/7 and do not have a chance to complain. Most challenging research is happening on ***energy consumption levels of data centers***. The Environmental Protection Agency (EPA) study showed that energy usage doubled from 2006 (61 billion kWh) to 2011. In 2009, data centers used 2% of worldwide power usage at expenditure of US $30 billion. Gartner had forecast cloud related expenditure for 2012 to be at $106.4 billion, a 12.7% increase from 2011, and revenue is forecast to change from $163 billion in 2011 to $240 billion in 2016. The IT sector contributed to 2% of carbon

Case id	For Memory		
	Error in User Memory estimate	Error in prediction by LWR	% reduction in error
For Uid= 148	69.5218	56.2085	21.18
For Uid= 169	173.6548	104.8720	49.39
For Uid= 412	209.9911	61.72	70.61

Fig. 7.16 Mean absolute percentage error and reduction in error for Memory estimate per job by using LWR

Paired Samples Test									
		Paired Differences							
		Mean	Std. Deviation	Std. Error Mean	95% Confidence Interval of the Difference		t	df	Sig. (2-tailed)
					Lower	Upper			
Pair 1	CPUrate_mean - DWA	.000919	.006860	.001194	-.00151	.003351	.769	32	.447
Pair 2	CPUrate_mean - LWR	.00285949	.00668205	.00116320	.00049014	.00522885	2.458	32	.020

Fig. 7.17 Result of paired ‘t’ test showing the accuracy of DWA for user uid 148

Paired Samples Test									
		Paired Differences							
		Mean	Std. Deviation	Std. Error Mean	95% Confidence Interval of the Difference		t	df	Sig. (2-tailed)
					Lower	Upper			
Pair 1	assignedmemoryusage_mean - MemDWA	-.00037	.013945	.002428	-.00532	.004574	-.153	32	.880
Pair 2	assignedmemoryusage_mean - MemLWR	.001591	.013620	.002371	-.00324	.006421	.671	32	.507

Fig. 7.18 Result of paired ‘t’ test showing accuracy of Memory prediction using DWA and LWR for user uid 148

dioxide worldwide in 2005, and it is growing by 6% per year. Nearly 80% of energy costs can be reduced by effective measures on various devices in a data center and 47 M metric tons of carbon dioxide emissions per year can be reduced. Google is the first major company that has revealed its power consumption information. Google uses 260 million watts continuously across the globe as mentioned in their blog-Google Green. Basic power consumption models for servers, storage devices, and network equipment in a data center is discussed in [38].

The power consumed on a data center when different workloads are run is described in [59]. In this work, the authors have clearly shown that for compute intensive workloads, CPU consumes large percentage of total machine power even while allowing DVFS (Dynamic CPU voltage and frequency scaling) that enable large reductions relative to total power. Furthermore, best practice techniques such as effective use of water available for cooling and extensive monitoring are now most commonly used approaches in large-scale Data centers [42]. A request routing framework FORTE has been proposed in [42]. They try to manage access latency within power costs with least possible carbon footprint. They are able to show reduction in carbon footprint, when Akamai use their approach by about 25% over 3 years, compared to carbon oblivious upgrade algorithm.

This makes it necessary to look at the various possible solutions.

7.5.2 Existing Approaches

Data center energy optimization is difficult because of issues like workload's requirement of specific types of servers, resource requirements for specific types of applications, cooling schedule, and request patterns. Data center energy-efficiency controllers for various components have to be addressed separately. Extra resources are deployed by the cloud provider to ensure 99.9% availability of services. Without bargain on Quality of service (QoS) and Service Level Agreements (SLAs), resources need to be managed carefully on the data center according to the workload requirements of cloud.

Software approach: specific approaches of optimizing resource use are being explored as below.

1. Live migration of VMs for improving energy-efficiency: This can lead to dynamic VM consolidation on the minimal number of physical hosts according to their current resource requirements. Placement of new requested VM or selected VMs for live migration is important issue in virtualized cloud and should be performed while ensuring QOS (Quality of Service), SLA (Service Level Agreement) and also preventing increase of total energy consumption by physical nodes in a data center. Various works in this area [60–62] have worked on energy-efficient live migration of VMs.
2. Reducing resource requests to ensure lower power consumption: Power consumption is proportional to CPU usage in a server. Well described in [4], we can use this opportunity to enhance energy-efficiency by reducing the unused CPU units for each user job.

7.5.3 Proposed Energy Conservation

Proposed approach works by optimization of energy consumption by **minimizing the number of resources being used** on active servers. We need to focus on the study of efficient resource utilization to minimize data center energy consumption by reducing the compute resources used. As experimentally shown for the Google trace data in [4], energy consumed by a machine is proportional to its CPU requests. **The more the CPU requests, the larger the frequency scaling factor, and this results in the cubic increment in power consumption.** Similarly in [63], the authors show that energy consumption per transaction is sensitive to variations in CPU utilization. Recent studies [59] and [64] show that between power consumption and CPU utilization a linear relationship exists. This forms the basis of proposed energy saving model. Hence, how energy can be saved directly by saving resource units is shown.

The selected trace, Google Cluster data, is a trace of users resource requests for Google compute cluster. The Google compute Engine has mentioned the use of Google Compute Units as the basic units that are made available to customers based on the number of units requested. The machines that are made available to users by various providers like Google Cloud Platform—Google Compute Engine—available at http://cloud.google.com/compute/docs. are quite variable. For example, Standard machine types, High memory machine types, High CPU machine types. Requests for these machines compute units, the virtual CPUs are actually implemented on Intel Sandy Bridge, Intel IVY Bridge, Intel Haswell machines. Google Compute Engine unit is a unit of CPU capacity that is used to describe the compute capability of machine types. Google has chosen 2.75 GCEUs to represent the minimum computational capacity of one virtual CPU (a hardware hyper-thread) on Sandy Bridge, Ivy Bridge, or Haswell platforms. Hence, for the two processors,Intel Xeon E5 2687 and Intel Xeon E5 2697 Energy saved is calculated based on idle Thermal Dissipation and Maximum energy consumed.

7.5.4 Energy Savings

Energy saved by using Resource prediction by DWA and Resource prediction by LWR are run for Intel Xeon E5 2687 and Intel Xeon E5 2697. The results show Percentage Energy saved **per job** for each of the users with uid 148, 169, and 412. The Percentage of Total energy saved **per user** is also computed and included in the results below.

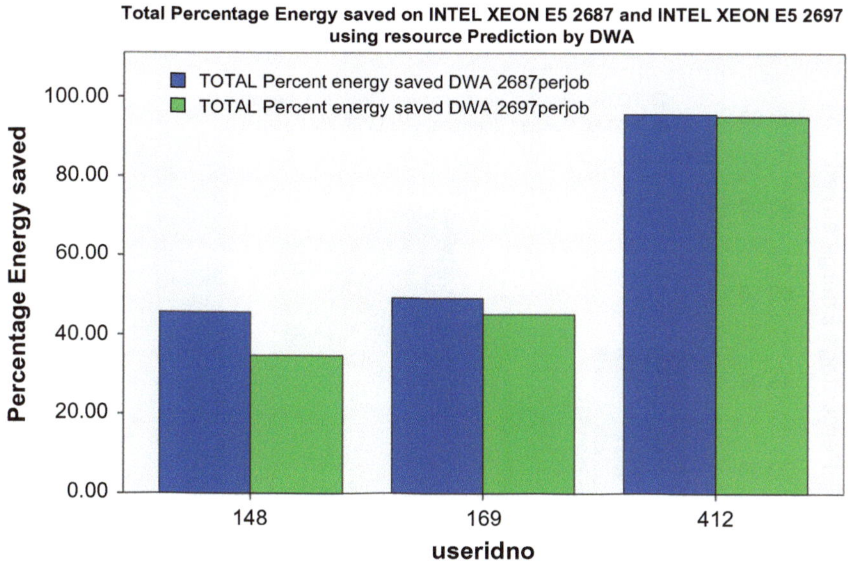

Fig. 7.19 Percentage of Total Energy saved **per user** (sum of energy savings of all jobs of each user) On Intel XEON E5 2687 and Intel Xeon E5 2697 due to resource requirement prediction by DWA

7.5.4.1 Energy Saved by Using DWA

To show the energy that is saved because of resource requirement prediction using prediction approach—DWA, bar graphs are used. These graphs clearly indicate for each job and for each user the energy that can be saved expressed as a percentage of the energy that is consumed without any prediction approach.

The bar graphs of Fig. 7.19 represent the Percentage of Total Energy saved **per user** by Resource requirement prediction by use of DWA on INTEL Xeon E5 2687 and INTEL Xeon E5 2697.

7.5.4.2 Energy Saved by Using LWR

To show the energy that is saved because of resource requirement prediction using prediction approach—LWR, bar graphs are used. These graphs clearly indicate for each job and for each user the energy that can be saved expressed as a percentage of the energy that is consumed without any prediction approach.

The bar graphs of Fig. 7.20 show the Percentage of Total Energy saved **PER USER** by Resource requirement prediction by use of LWR on INTEL Xeon E5 2687 and INTEL Xeon E5 2697.

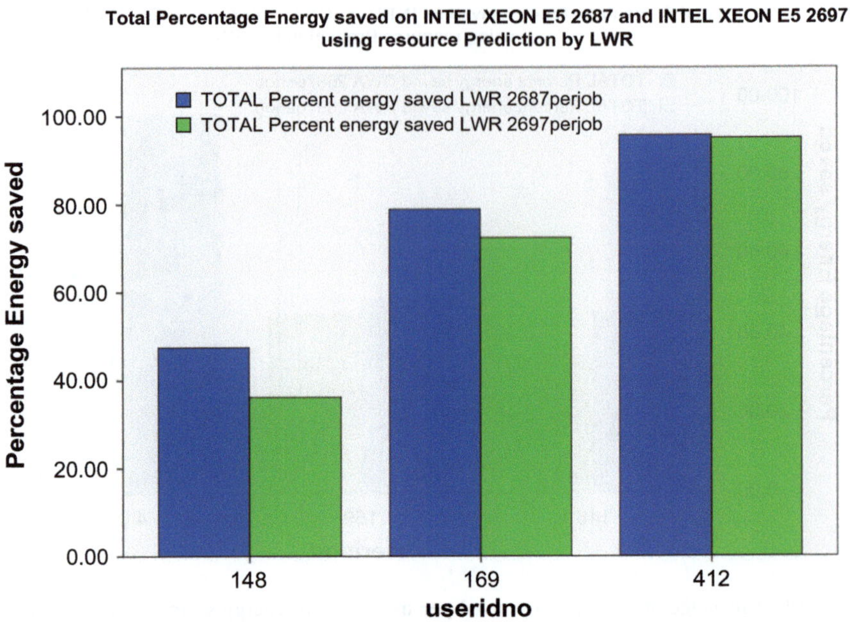

Fig. 7.20 Percentage of Total Energy saved **per user** On Intel XEON E5 2687 and Intel Xeon E5 2697 due to resource requirement prediction by LWR

7.5.5 Analysis of Energy Savings Obtained

The results are available in comprehensive comparative tables.

1. Energy Savings obtained
   (a) Energy savings in Intel Xeon E5 2687
      - Saving in Energy by prediction using DWA Saving in Energy by prediction using DWA for Intel 2687 is available in a table as shown in Fig. 7.21. The table also shows percentage reduction in energy wastage and percentage saving of energy.
      - Saving in Energy by prediction using LWR Saving in Energy by prediction using LWR for Intel 2687 is available in a table as shown in Fig. 7.22.

   (b) Energy savings in Intel Xeon E5 2697
      - Saving in energy by prediction using DWA Saving in Energy by prediction using DWA for Intel 2697 is available in a table as shown in Fig. 7.23.
      - Saving of energy by prediction using LWR in 2697 Saving in Energy by prediction using LWR for Intel 2697 is available in a table as shown in Fig. 7.24.

FOR DWA on Intel Xeon E5 2687		Observed Variation as compared to Energy actually Used		Percentage reduction in variation	Saving of energy units	Percentage saving of energy
	Energy consumed with Adhoc requests (A3)	Energy used with Adhoc requests – Energy used with actual resource used (A1)	Energy used with DWA prediction – Energy used actually by user (A2)	((A1 – A2) /A1) * 100	Energy used with Adhoc Requests – Energy used with DWA prediction (A4)	(A4/A3) *100
UId= 148	305.57	131.11	-8.49	93.5168	139.61	45.69
Uid =169	1069.06	509.25	-17.06	96.6499	526.31	49.23
Uid = 412	13592.35	12590.75	-412.84	96.7210	13003.59	95.67

Fig. 7.21 Energy saved per job by using DWA in 2687

FOR LWR on Intel Xeon E5 2687		Observed Variation as compared to Energy actually Used		Percentage reduction in variation	Saving of energy units	Percentage saving of energy
	Energy consumed with Adhoc requests (A3)	Energy used with Adhoc requests – Energy used with actual resource used (A1)	Energy used with LWR prediction – Energy used actually by user (A2)	((A1 – A2) /A1) * 100	Energy used with Adhoc Requests – Energy used with LWR prediction (A4)	(A3 /A4) *100
UId= 148	305.57	131.11	-13.97	68.1449	145.08	47.48
Uid =169	1069.06	509.25	-334.63	34.2896	843.88	78.94
Uid = 412	13592.35	12590.75	-419.66	96.6669	13010.41	95.72

Fig. 7.22 Energy saved per job by using LWR in 2687

7.6 Contributions

Proposed work uses Enhanced Instance Based Learning approach for resource requirements prediction and runs this approach with actual Cloud trace data. The energy saving that would be achieved with predicted resources as compared to existing user requests for resources is shown.

FOR DWA on Intel Xeon E5 2697		Observed Variation as compared to Energy actually Used		Percentage reduction in variation	Saving of energy units	Percentage saving of energy
	Energy consumed with Adhoc requests (A3)	Energy used with Adhoc requests – Energy used with actual resource used (A1)	Energy used with DWA prediction – Energy used actually by user (A2)	((A1 – A2) /A1) * 100	Energy used with Adhoc Requests – Energy used with DWA prediction	(A4/A3) *100
UId= 148	212.36	69.41	-4.50	93.5167	73.91	34.80
Uid =169	616.56	269.60	-9.03	96.6505	278.64	45.19
Uid = 412	7246.48	6665.69	-218.56	96.7211	6884.25	95.00

Fig. 7.23 Energy saved per job by using DWA in 2697

FOR LWR on Intel Xeon E5 2697		Observed Variation as compared to Energy actually Used		Percentage reduction in variation	Saving of energy units	Percentage saving of energy
	Energy consumed with Adhoc requests (A3)	Energy used with Adhoc requests – Energy used with actual resource used (A1)	Energy used with LWR prediction – Energy used actually by user (A2)	((A1 – A2) /A1) * 100	Energy used with Adhoc Requests – Energy used with LWR prediction (A4)	(A4/A3) *100
UId= 148	212.36	69.41	-7.40	89.3387	76.81	36.14
Uid =169	616.56	296.60	-177.16	40.2697	446.76	72.46
Uid = 412	7246.48	6665.69	-222.17	96.6669	6887.87	95.05

Fig. 7.24 Energy saved per job by using LWR in 2697

Results obtained from proposed predictive approach shows Resource saving (compute and Memory), energy saved, accuracy of prediction, and data size used by prediction.

There is a considerable improvement over existing ad-hoc approach by the users. Specific contributions can be enlisted as:

1. When DWA and LWR are employed for prediction, the savings of resource units per job of a user are shown in results. Resource units saved per job as percentage of resource saved compared to what was actually getting wasted by ad-hoc method of request for resource is found to be on average—90.68% saving for CPU estimates using DWA, 73.44% saving for CPU estimates using LWR, and 73.28% saving for Memory estimates using DWA, 88.47% saving for Memory estimates using LWR.
2. The percentage reduction in resource estimation error is on average—95% for CPU estimates using DWA, 92.5% for CPU estimates using LWR and 67.13% for Memory estimates using DWA, 47% for Memory estimates using LWR.
3. The resource units saved by DWA and LWR Compute units prediction approach also helps in computing Energy saved on INTEL Xeon E5 2687 and INTEL Xeon E5 2697. Energy saving computation for various jobs shows average saving of 63% (detailed energy saving is described in Sect. 7.5.4).
4. The proposed approaches for prediction use small amount of data for processing to arrive at the result. Use of *K means* achieves this objective, as mentioned in subsection. Hence, it requires less time to process the prediction requirements. This is an important consideration for Cloud systems where dynamic reconfiguration based on users requests and actual use of resources will play a crucial role in successfully incorporating this prediction approach into real cloud provider system in the future.

Finally, the results of prediction of resources have been executed and tested with real time data trace of Cloud. Hence, this is an efficient approach to Capacity Planning for large data centers. Further, adoption of this approach into tools that can display to user information on resource availability, reservation and billing information will be extremely useful.

7.7 Conclusion and Future Work

7.7.1 Conclusion

Resource provisioning is a challenging area of research. Saving resources on cloud by the resource providers and paying for only those resources that will actually be used, rather than paying for resources reserved by the user is the motivation for this work. This is business advantage if this work can be adopted by cloud providers.

There is no direct relationship between the users requests for resources and actual usage values. Also, there is no pattern in his resource usage data. This made use of various existing approaches time series, Queuing models, Neural networks, etc., infeasible for the cloud environment. The real cloud usage is much different from any synthetic workload that can be generated. Proposed Cloud Resource usage prediction method has been evaluated using Google Cluster Data. Using the proposed Enhanced Instance Based Learning (EIBL), the saving of resource units per collection of machines per hour is quite significant. This also translates to significant saving of energy units per user jobs. This enables better and efficient utilization of the resources by the service provider. This proposed approach when scaled and worked in real cloud system will be extremely beneficial to both—the users and service providers.Overall, this is a substantial contribution to existing knowledge on much needed effective prediction techniques for Cloud.

7.7.2 Future Work

Capacity planning is very useful consequence of this work. Machine consolidation based on efficient resource usage, available from our prediction and load balancing, efficient resource utilization is possible. With lesser use of machines, energy can be saved, leading to Green computing. Efficient Scaling is now possible because resource requirements are predicted beforehand. **Therefore, applications that require instant and variable resources can easily adopt this EIBL approach to predictive scaling and efficient energy saving**. Further, adoption of this approach into tools that can display to user information on resource availability, reservation and billing information will be extremely useful. **As an extension, the energy saved by use of this work, hence, green cover saved can be explicitly computed.**

References

1. Reiss C, Tumanov A, Ganger GR, Katz RH, Kozuch MA. Heterogeneity and dynamicity of clouds at scale: Google trace analysis. In: Proceedings of the third ACM symposium on cloud computing. ACM; 2012. p. 7.
2. Hamilton J. Cooperative expendable micro-slice servers (cems): low cost, low power servers for internet-scale services. In: Conference on innovative data systems research (CIDR'09) (Jan 2009); 2009.
3. Garraghan P, Townend P, Xu J. An analysis of the server characteristics and resource utilization in google cloud. In: Cloud engineering (IC2E), 2013 IEEE international conference on. IEEE; 2013. p. 124–131.
4. Dong Z, Zhuang W, Rojas-Cessa R. Energy-aware scheduling schemes for cloud data centers on google trace data. In: Green communications (OnlineGreencomm), 2014 IEEE online conference on. IEEE; 2014. p. 1–6.
5. Cloud AEC. Web page at http://aws.amazon.com/ec2. Date of last access: 14 Sept 2010.
6. Bryant R, Tumanov A, Irzak O, Scannell A, Joshi K, Hiltunen M, Lagar-Cavilla A, De Lara E. Kaleidoscope: cloud microelasticity via vm state coloring. In: Proceedings of the sixth conference on computer systems. ACM; 2011. p. 273–286.
7. Nguyen H, Shen Z, Gu X, Subbiah S, Wilkes J. Agile: elastic distributed resource scaling for infrastructure as a service. In: Proceedings of the USENIX international conference on automated computing (ICAC'13). San Jose, CA; 2013.
8. Manage massive amounts of data, fast, without losing sleep. 2016. The Apache Software Foundation Retrived July 10th, 2017 from http://cassandra.apache.org
9. Cloud can be complex. Make it simple with RightScale. 2005–2017. RightScale Inc. Retrived July 10th, 2017 from http://www.rightscale.com

10. Qureshi A, Weber R, Balakrishnan H, Guttag J, Maggs B. Cutting the electric bill for internet-scale systems. ACM SIGCOMM Comput. Commun Rev. 2009;39(4):123–134. ACM.
11. Ali-Eldin A, Kihl M, Tordsson J, Elmroth E. Efficient provisioning of bursty scientific workloads on the cloud using adaptive elasticity control. In: Proceedings of the third workshop on scientific cloud computing date. ACM; 2012. p. 31–40.
12. Ali-Eldin A, Tordsson J, Elmroth E., Kihl M. Workload classification for efficient autoscaling of cloud resources. Technical Report; 2005 [Online]. http://www8.cs.umu.se/research/uminf/reports/2013/013/part1.pdf. Tech. Rep., 2013.
13. Addis B, Ardagna D, Panicucci B, Zhang L. Autonomic management of cloud service centers with availability guarantees. In: Cloud computing (CLOUD), 2010 IEEE third international conference on. IEEE; 2010. p. 220–227.
14. Beloglazov A, Buyya R. Energy efficient resource management in virtualized cloud data centers. In: Proceedings of the 2010 tenth IEEE/ACM international conference on cluster, cloud and grid computing. IEEE Computer Society; 2010. p. 826–831.
15. Ferretti S, Ghini V, Panzieri F, Pellegrini M, Turrini E. Qos–aware clouds. In: Cloud computing (CLOUD), 2010 IEEE third international conference on. IEEE; 2010. p. 321–328.
16. Teng F, Magoules F. Resource pricing and equilibrium allocation policy in cloud computing. In: Computer and information technology (CIT), 2010 IEEE tenth international conference on. IEEE; 2010. p. 195–202.
17. Yazir YO, Matthews C, Farahbod R, Neville S, Guitouni A, Ganti S, Coady Y. Dynamic resource allocation in computing clouds using distributed multiple criteria decision analysis. In: Cloud computing (CLOUD), 2010 IEEE third international conference on. IEEE; 2010. p. 91–98.
18. Liu S, Quan G, Ren S. On-line scheduling of real-time services for cloud computing. In: Services (SERVICES-1), 2010 sixth world congress on. IEEE; 2010. p. 459–464.
19. Van HN, Tran FD, Menaud J-M. Sla-aware virtual resource management for cloud infrastructures. In: Computer and information technology, 2009. CIT'09. Ninth IEEE international conference on, vol. 1. IEEE; 2009. p. 357–362.
20. Chaisiri S, Lee B-S, Niyato D. Optimal virtual machine placement across multiple cloud providers. In: Services computing conference, 2009. APSCC 2009. IEEE Asia-Pacific. IEEE; 2009. p. 103–110.
21. Van HN, Tran FD, Menaud J-M. Performance and power management for cloud infrastructures. In: Cloud computing (CLOUD), 2010 IEEE third international conference on. IEEE; 2010. p. 329–336.
22. Wang W, Chen H, Chen X. An availability aware virtual machine placement approach for dynamic scaling of cloud applications. In: Ubiquitous intelligence & computing and ninth international conference on autonomic & trusted computing (UIC/ATC), 2012 ninth international conference on. IEEE; 2012. p. 509–516.
23. Redkar T, Guidici T. Windows azure platform. Berlin Heidelberg: Springer; 2009.
24. Singhai A, Lim S, Radia SR. The scalr framework for internet services. In: Proceedings of the twenty-eighth fault-tolerant computing symposium (FTCS-28); 1998. page (to appear).
25. Mao M, Humphrey M. Autoscaling to minimize cost and meet application deadlines in cloud workflows. In: Proceedings of 2011 international conference for high performance computing, networking, storage and analysis. ACM; 2011. p. 49.
26. Roy N, Dubey A, Gokhale A. Efficient autoscaling in the cloud using predictive models for workload forecasting. In: Cloud computing (CLOUD), 2011 IEEE international conference on. IEEE; 2011. p. 500–507.
27. Brockwell PJ. Introduction to time series and forecasting, vol. 1. Abingdon: Taylor & Francis; 2002.
28. Chatfield C. The analysis of time series: an introduction. Boca Raton: CRC Press; 2013.
29. Dinda PA, O'Hallaron DR. Host load prediction using linear models. Clust Comput. 2000;3(4):265–280.

30. Gmach D, Rolia J, Cherkasova L, Kemper A. Capacity management and demand prediction for next generation data centers. In: Web services, 2007. ICWS 2007. IEEE international conference on. IEEE; 2007. p. 43–50.
31. Huang J, Li C, Yu J. Resource prediction based on double exponential smoothing in cloud computing. In: Consumer electronics, communications and networks (CECNet), 2012 second international conference on. IEEE; 2012. p. 2056–2060.
32. Geweke J, Whiteman C. Bayesian forecasting. In: Handbook of economic forecasting, vol. 1. Amsterdam: Elsevier; 2006. p. 3–80.
33. Di S, Kondo D, Cirne W. Host load prediction in a google compute cloud with a Bayesian model. In: Proceedings of the international conference on high performance computing, networking, storage and analysis. Washington, DC: IEEE Computer Society Press; 2012. p. 21.
34. Mohammadi K, Eslami H, Dardashti SD, et al. Comparison of regression, ARIMA and ANN models for reservoir inflow forecasting using snowmelt equivalent (a case study of Karaj). J Agric Sci Technol. 2005;7:17–30.
35. Gong Z, Gu X, Wilkes J. Press: predictive elastic resource scaling for cloud systems. In: Network and service management (CNSM), 2010 international conference on. IEEE; 2010. pp. 9–16.
36. Jiang Y, Perng C-s, Li T, Chang R. Asap: A self-adaptive prediction system for instant cloud resource demand provisioning. In: Data mining (ICDM), 2011 IEEE eleventh international conference on. IEEE; 2011. p. 1104–1109.
37. Miu T, Missier P. Predicting the execution time of workflow blocks based on their input features. Computing Science, Newcastle University, 2013.
38. Quan DM, Basmadjian R, De Meer H, Lent R, Mahmoodi T, Sannelli D, et al. Energy efficient resource allocation strategy for cloud data centres. In: Computer and information sciences II. London: Springer; 2012. p. 133–141.
39. Rodero I, Jaramillo J, Quiroz A, Parashar M, Guim F, Poole S. Energy-efficient application-aware online provisioning for virtualized clouds and data centers. In: Green computing conference, 2010 international. IEEE; 2010. p. 31–45.
40. Abdelsalam HS, Maly K, Mukkamala R, Zubair M, Kaminsky D. Analysis of energy efficiency in clouds. In: Future computing, service computation, cognitive, adaptive, content, patterns, 2009. COMPUTATIONWORLD'09. Computation World. IEEE; 2009. p. 416–421.
41. Younge AJ, Von Laszewski G, Wang L, Lopez-Alarcon S, Carithers W. Efficient resource management for cloud computing environments. In: Green computing conference, 2010 international. IEEE; 2010. p. 357–364
42. Gao PX, Curtis AR, Wong B, Keshav S. It's not easy being green. ACM SIGCOMM Comput Commun Rev. 2012;42(4):211–222.
43. Wang D, Ren C, Sivasubramaniam A, Urgaonkar B, Fathy H. Energy storage in datacenters: what, where, and how much? ACM SIGMETRICS Perform Eval Rev. 2012;40(1):187–198. ACM.
44. Kurpicz M, Sobe A, Felber P. Using power measurements as a basis for workload placement in heterogeneous multi-cloud environments. In: Proceedings of the second international workshop on crosscloud systems. ACM; 2014. p. 6.
45. Chen F, Grundy J, Schneider J-G, Yang Y, He Q. Automated analysis of performance and energy consumption for cloud applications. In: Proceedings of the fifth ACM/SPEC international conference on performance engineering. ACM; 2014. p. 39–50
46. Mishra AK, Hellerstein JL, Cirne W, Das CR. Towards characterizing cloud backend workloads: insights from google compute clusters. ACM SIGMETRICS Perform Eval Rev. 2010;37(4):34–41.
47. Aggarwal S, Phadke S, Bhandarkar M. Characterization of Hadoop jobs using unsupervised learning. In: Cloud computing technology and science (CloudCom), 2010 IEEE second international conference on. IEEE; 2010. p. 748–753.
48. Tan J, Dube P, Meng X, Zhang L. Exploiting resource usage patterns for better utilization prediction. In: Distributed computing systems workshops (ICDCSW), 2011 thirty-first international conference on. IEEE; 2011. p. 14–19.

49. Zhang Q, Hellerstein JL, Boutaba R. Characterizing task usage shapes in google's compute clusters. In: Large scale distributed systems and middleware workshop (LADIS'11); 2011.
50. Sharma B, Chudnovsky V, Hellerstein JL, Rifaat R, Das CR. Modeling and synthesizing task placement constraints in google compute clusters. In: Proceedings of the second ACM symposium on cloud computing. ACM; 2011. p. 3.
51. Di S, Kondo D, Cirne W. Characterization and comparison of cloud versus grid workloads. In: Cluster computing (CLUSTER), 2012 IEEE international conference on. IEEE; 2012. p. 230–238.
52. Moreno IS, Garraghan P, Townend P, Xu J. An approach for characterizing workloads in google cloud to derive realistic resource utilization models. In: Service oriented system engineering (SOSE), 2013 IEEE seventh international symposium on. IEEE; 2013. p. 49–60.
53. Amoretti M, Lafuente AL, Sebastio S. A cooperative approach for distributed task execution in autonomic clouds. In: Parallel, distributed and network-based processing (PDP), 2013 twenty-first Euromicro international conference on. IEEE; 2013. p. 274–281.
54. Hellerstein JL, Cirne W, Wilkes J. Google cluster data. Google research blog, Jan, 2010.
55. Reiss C, Wilkes J, Hellerstein JL. Google cluster-usage traces: format+ schema, Technical report. Mountain View, CA, USA: Google Inc.; 2011.
56. Jiang Y., Perng C-s, Li T, Chang R. Self-adaptive cloud capacity planning. In: Services computing (SCC), 2012 IEEE ninth international conference on. IEEE; 2012. p. 73–80.
57. Kavulya S, Tan J, Gandhi R, Narasimhan P. An analysis of traces from a production mapreduce cluster. In: Cluster, cloud and grid computing (CCGrid), 2010 tenth IEEE/ACM international conference on. IEEE; 2010. p. 94–103.
58. Buana PW, Darma IKG. Combination of k-nearest neighbor and k-means based on term re-weighting for classify Indonesian news. Int J Comput Appl. 2012;50(11): 37–42.
59. Fan X, Weber W-D, Barroso LA. Power provisioning for a warehouse-sized computer. ACM SIGARCH Comput. Archit News. 2007;35(2):13–23. ACM.
60. Zheng J, Ng TSE, Sripanidkulchai K. Workload-aware live storage migration for clouds. ACM SIGPLAN Notices. 2011;46(7):133–144. ACM.
61. Daniel S, Kwon M. Prediction-based virtual instance migration for balanced workload in the cloud datacenters. Rochester: Rochester Institute of Technology Scholar Works; 2011.
62. Abdulmohson A, Pelluri S, Sirandas R. Energy efficient load balancing of virtual machines in cloud environments. International Journal of Cloud-Computing and Super-Computing. 2015;2(1):21–34.
63. Srikantaiah S, Kansal A, Zhao F. Energy aware consolidation for cloud computing. In: Proceedings of the 2008 conference on power aware computing and systems, vol. 10. San Diego, CA; 2008.
64. Kusic D, Kephart JO, Hanson JE, Kandasamy N, Jiang G. Power and performance management of virtualized computing environments via lookahead control. Clust Comput. 2009;12(1):1–15.

Chapter 8
Short-Term Prediction Model to Maximize Renewable Energy Usage in Cloud Data Centers

Atefeh Khosravi and Rajkumar Buyya

8.1 Introduction

Cloud computing is a paradigm focused on the realization and long held dream of delivering computing as a utility [1]. It enables businesses and developers access to hardware resources and infrastructure anytime and anywhere they want. Nowadays, the number of individuals and organizations shifting their workload to cloud data centers is growing more than ever. Cloud services are delivered by data center sites each containing tens of thousands of servers, which are distributed across geographical locations. The geographical diversity of computing resources brings several benefits, such as high availability, effective disaster recovery, uniform access to users in different regions, and access to different energy sources.

Over the recent years the use of services offered by cloud computing systems has been increased and different definitions for cloud computing have been proposed. According to the definition by the National Institute of Standards and Technology (NIST) [2]: "Cloud computing is a model for enabling ubiquitous, convenient, on-demand network access to a shared pool of configurable computing resources (e.g., networks, servers, storage, applications, and services) that can be rapidly provisioned and released with minimal management effort or service provider interaction."

Cloud addresses the issue of under provisioning of resources for a running service and lose the potential users at the peak times or even over provisioning of resources that leads to wastage of capital costs. This definition highlights a major feature for cloud computing that is called elasticity of resources. By delivering computing as a utility to users and providing the resources based on the users'

A. Khosravi (✉) • R. Buyya
Cloud Computing and Distributed Systems (CLOUDS) Laboratory, School of Computing and Information Systems, The University of Melbourne, Parkville, VIC, Australia
e-mail: atefehk@student.unimelb.edu.au; rbuyya@unimelb.edu.au

W. Rivera (ed.), *Sustainable Cloud and Energy Services*,
DOI 10.1007/978-3-319-62238-5_8

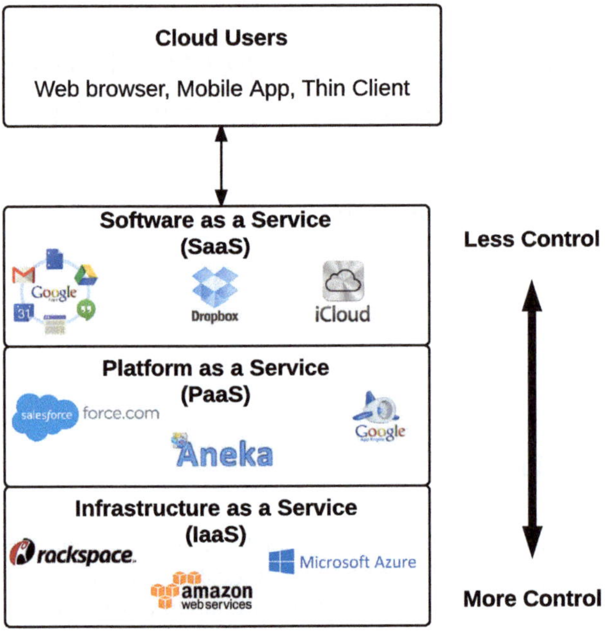

Fig. 8.1 Cloud computing services

request, the users will be charged on a pay-as-you-go manner, such as other utility pricing models (e.g., electricity and water). In other words, users need not pay any upfront cost and the billing will be based on the usage (e.g. hourly) of the cloud resources.

Cloud delivers three main services to users as shown in Fig. 8.1 and discussed in the following.

- Software as a Service: At the highest level there is Software as a Service (SaaS). SaaS service model, which is an old idea of cloud computing delivers on-demand software to users. Google Apps [3] and Salesforce [4] are examples of services offered in SaaS model. In this model, the control, support, and maintenance of the hardware, platform, and software of the cloud environment is shifted from the end-user to the cloud provider.
- Platform as a Service: Platform as a Service (PaaS) provides computing platform with pre-installed operating system, in order to enable the developers create their own software. By using PaaS, the developers need not concern about the underlying hardware and the operating system. Users can have scalable resources anytime and anywhere. Google App Engine [5], Microsoft Azure [6], and Manjrasoft Aneka [7] are examples of PaaS environment.
- Infrastructure as a Service: Infrastructure as a Service (IaaS) located at the lowest layer of the cloud service stack offers computing physical resources such as servers, storage, hardware, networking, and virtual machines (VMs) to

users. In this model, users have control over the operating system, storage, and applications while they need not manage the underlying infrastructure. Amazon EC2 [8], Google Cloud [9], and Rackspace [10] are some of the well-known IaaS providers.

Services offered by cloud computing are delivered by data centers distributed across the world. One major issue with these data centers is that they are energy intensive, which makes them responsible for 2% of the world's total CO_2 emission [11]. To overcome the problem of high energy consumption and environmental concerns due to the high CO_2 emission of energy sources, there are possible solutions such as improving the data center's efficiency or replacing the polluting (brown) energy sources with clean energy sources. By making data centers aware of energy sources and better utilizing renewable energy, cloud providers are able to reduce the energy consumption and carbon footprint significantly [12].

Further in this section, we elaborate more on our motivation, which is one of the biggest challenges a cloud provider faces, high energy consumption and carbon footprint, and the need to better utilize renewable energy sources.

8.1.1 Motivation

Data centers are the backbone of the Internet that consist of thousands of servers. They are one of the fastest growing industries that offer different types of services to users around the world. However, data centers are known to consume huge amount of electricity. According to a report by NRDC [13], US data centers in 2013 alone consumed 91 billion kWh of electricity. This is equivalent to 2-year power consumption of New York City's households and by 2020 is estimated to increase to 140 billion kWh. This could be equivalent to nearly 150 million tons of carbon pollution. Therefore, many cloud service providers focused on reducing their reliance on electricity driven from fossil fuels and transition to renewable energy sources.

Recently, large cloud providers started building their on-site renewable energy sources. Companies such as Amazon [14], Facebook [15, 16], Google [17], and Microsoft [18] all have their own on-site solar/wind farms. Renewable energy sources have intermittent nature. This means that their availability changes during the day and based on time of the year. However, since all the large cloud providers have geographically distributed data center sites, as depicted in Fig. 8.2, they can benefit from this location diversity. This helps them to migrate the user requests (e.g., VMs) in the absence of renewable energy in a data center to a site with excess renewable energy.

Since, most sources of renewable energy have intermittent nature knowing the future level of energy helps the cloud provider to make informed decision on when to migrate the VMs to maximize renewable energy usage. The cloud provider can benefit from short-term prediction of renewable energy to perform

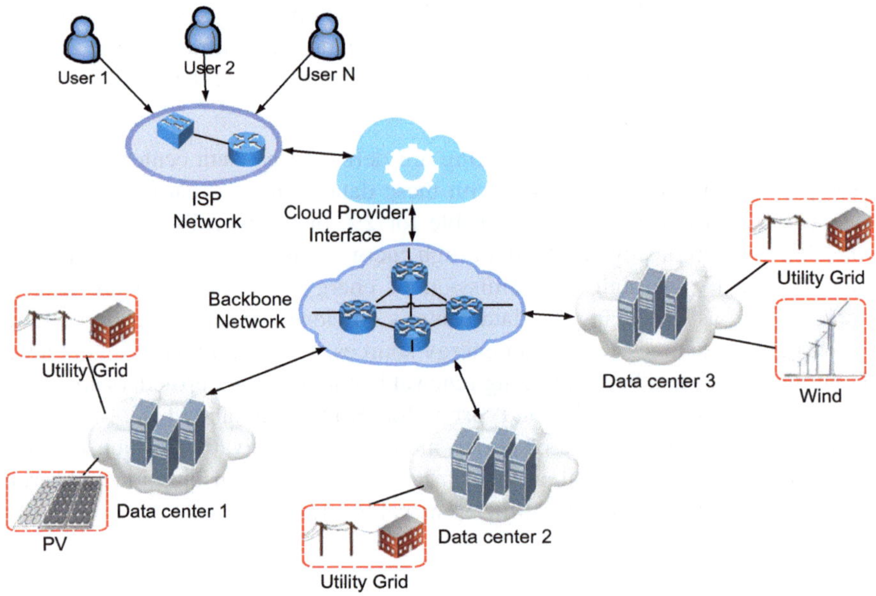

Fig. 8.2 Cloud provider with distributed data center sites with different energy sources

future-aware online algorithms to migrate the VMs, as it has been stated in our previous work [19]. This helps the provider to increase the performance of the online algorithms close to the optimal offline, which has full knowledge of the future level of renewable energy.

In this chapter, we propose a short-term prediction model based on the Gaussian mixture model [20]. The proposed model predicts renewable energy level for many-steps ahead into the future. A primary requirement to perform prediction is knowing the current and previous states of the renewable energy levels, since the future level can be inferred from current and previous states and their correlation. The GMM model uses history data to train itself. We use renewable energy measurements reported by NREL [21], that have been used in our previous work [19] as real meteorological data, as history and test data in our experiments. Moreover, we verified the accuracy of the proposed prediction model using workload demand collected from AWS biggest region, US East, Virginia. However, due to the confidentiality of that data set, we only rely on the analysis carried out using renewable energy traces collected from NREL.

The rest of the chapter is organized as follows: Sect. 8.2 describes the prediction model objective. The formulation and component estimation of the prediction model is explained in Sect. 8.3. Section 8.4 elaborates on the required steps to construct the model. The approaches and methodologies to train the history data are explained in Sect. 8.5. Experiment results are presented in Sects. 8.6 and 8.7 provides a summary of the chapter.

8.2 Prediction Model Objective

Energy production at a data center within time period $[1, T]$ is time-series data and can be shown as $\mathbf{y} = [y_1, y_2, \ldots, y_T]^T$, where y_t is the energy production at time t. We show the predicted renewable energy production in a data center at time t as $\hat{y}_t$. The closer the predicted energy $\hat{y}_t$ is to the observed production energy y_t, the more accurate the prediction.

Therefore, our objective is to minimize the prediction error over time interval $[t_1, t_2]$ where $t_1 \leq t_2$, and is stated as follows:

$$\begin{aligned} \underset{\hat{y}_t}{\text{minimize}} \quad & \sum_{t \in [t_1, t_2]} e[(\hat{y}_t - y_t)] \,, \\ \text{subjectto} \quad & \hat{y}_t \geq 0 \,, \\ \text{and} \quad & \text{predictionModelCost} \leq \text{ThresholdCost} \,. \end{aligned} \tag{8.1}$$

The first constraint guarantees the predicted energy production always has non-negative value. Finally, the second constraint guarantees the computation cost of running the prediction, in terms of running time, CPU, and memory usage over a certain time period will not exceed a predetermined threshold.

8.3 Prediction Model Formulation

We use the current and previous states of the energy production to perform prediction. The next state of energy production has strong but not deterministic relationship with the current and previous states. This relationship could be shown as a conditional probability. If we denote the current state of the energy production as y_t, then the probability of the next state can be denoted as:

$$p(y_{t+1} | y_t, y_{t-1}, \ldots, y_{t-N+1}) \,, \tag{8.2}$$

where N is considered as the number of previous states taken into account for the prediction. For the sake of simplicity, we show the previous states considered in the prediction as $\mathbf{x} = [y_t, y_{t-1}, \ldots, y_{t-N+1}]^T$. Therefore, to obtain the energy production prediction we need to compute the following conditional estimation:

$$\hat{y}_{t+1} = \mathbb{E}[y_{t+1} | \mathbf{x}] \,. \tag{8.3}$$

8.3.1 Prediction Using Gaussian Mixture Model

To perform the prediction in near future using historical renewable energy production, we use Gaussian mixture models (GMM). In order to obtain the prediction value, first we need to compute $p(y_{t+1}|\mathbf{x})$. Since the aforementioned probability is unknown, we use GMM to approximate it, assuming it is a combination of multiple Gaussian components [20]. GMM is a powerful tool for data analysis and is characterized by M number of mixtures/components, each with a given mean $\boldsymbol{\mu}$, variance $\boldsymbol{\Sigma}$, and weight ω. The GMM probability density function can be written as follows:

$$p(\mathbf{x}|\Theta) = \sum_{j=1}^{M} \omega_j \mathcal{N}(\mathbf{x}; \boldsymbol{\mu}_j, \boldsymbol{\Sigma}_j)\,, \tag{8.4}$$

where

$$\begin{aligned} &\Theta = \{(\omega_1, \boldsymbol{\mu}_1, \boldsymbol{\Sigma}_1), (\omega_2, \boldsymbol{\mu}_2, \boldsymbol{\Sigma}_2), \ldots (\omega_M, \boldsymbol{\mu}_M, \boldsymbol{\Sigma}_M)\}\,, \\ &\sum_{j=1}^{M} \omega_j = 1\,, \\ &\mathcal{N}(\mathbf{x}, ; \boldsymbol{\mu}_j, \boldsymbol{\Sigma}_j) = \frac{1}{\boldsymbol{\Sigma}_j \sqrt{2\pi}} e^{-\frac{(\mathbf{x}-\boldsymbol{\mu}_j)^2}{2\boldsymbol{\Sigma}_j}}\,. \end{aligned} \tag{8.5}$$

GMM parameters, Θ, can be estimated using the expectation-maximization (EM) algorithm [22]. EM is the most popular approach being used and it iteratively optimizes the model using maximum likelihood maximization.

As we mentioned before, the next energy production value has a conditional probability with the current and previously observed production:

$$\begin{aligned} \hat{y} &= \mathbb{E}[y|\mathbf{x}] \\ &= \int y p(y|\mathbf{x}) dy\,. \end{aligned} \tag{8.6}$$

Since $p(y|\mathbf{x})$ in the Eq. (8.6) is not known, we use Bayes' Theorem for its estimation stated as follows:

$$p(y|x) = \frac{p(y, \mathbf{x})}{p(\mathbf{x})}\,, \tag{8.7}$$

and the joint probability distribution for y and $\mathbf{x}$, $p(y, \mathbf{x})$, could be derived using GMM. Therefore, Eq. (8.7) could be restated as:

$$
\begin{aligned}
p(y|\mathbf{x}) &= \frac{\sum_{i=1}^{M} \omega_i \mathcal{N}(\mathbf{x}; \boldsymbol{\mu}_{i\mathbf{x}^T}, \boldsymbol{\Sigma}_{i\mathbf{xx}}) \mathcal{N}(y; \boldsymbol{\mu}_{iy|\mathbf{x}^T} \boldsymbol{\Sigma}_{iy|\mathbf{x}})}{\sum_{j=1}^{M} \mathcal{N}(\mathbf{x}; \boldsymbol{\mu}_{j\mathbf{x}^T}, \boldsymbol{\Sigma}_{j\mathbf{xx}})} \\
&= \sum_{i=1}^{M} \beta_i \mathcal{N}(y; \boldsymbol{\mu}_{iy|\mathbf{x}^T}, \boldsymbol{\Sigma}_{iy|\mathbf{x}}) \,,
\end{aligned}
\tag{8.8}
$$

where

$$
\begin{aligned}
\beta_i &= \frac{\omega_i \mathcal{N}(\mathbf{x}; \boldsymbol{\mu}_{i\mathbf{x}^T}, \boldsymbol{\Sigma}_{i\mathbf{xx}})}{\sum_{j=1}^{M} \omega_j \mathcal{N}(\mathbf{x}; \boldsymbol{\mu}_{j\mathbf{x}^T}, \boldsymbol{\Sigma}_{j\mathbf{xx}})} \,, \\
\boldsymbol{\mu}_{iy|\mathbf{x}^T} &= \mu_{iy} - \boldsymbol{\Sigma}_{iy\mathbf{x}} \boldsymbol{\Sigma}_{i\mathbf{xx}}^{-1} (\boldsymbol{\mu}_{i\mathbf{x}^T} - \mathbf{x}) \,.
\end{aligned}
\tag{8.9}
$$

Finally, by substituting Eq. (8.8) into Eq. (8.6), we have:

$$
\begin{aligned}
\hat{y} &= \sum_{i=1}^{M} \beta_i \int y \mathcal{N}(y; \boldsymbol{\mu}_{iy|\mathbf{x}^T}, \boldsymbol{\Sigma}_{iy|\mathbf{x}^T}) dy \\
&= \sum_{i=1}^{M} \beta_i \boldsymbol{\mu}_{iy|\mathbf{x}^T} \,.
\end{aligned}
\tag{8.10}
$$

8.3.2 Optimal GMM Components Estimation

We use expectation maximization (EM) algorithm to estimate GMM parameters Θ. EM is an iterative method to find the maximum likelihood estimate (MLE) of the parameters. In order for EM to perform the two steps of expectation (E) and maximization (M), it needs to receive the number of GMM mixtures as an input.

There have been several studies and different methods to obtain the optimal number of mixtures and selecting the efficient model, rather than simply taking a random or educated guess. Bayesian information criterion (BIC) [23] is a criterion introduced for model selection and is penalized based on the model complexity. BIC maximizes the maximum likelihood function for each model. It is based on the increasing function of an error and the model with the lowest BIC, the more efficient in terms of predicting the demand.

8.4 Construction of Prediction Model

Figure 8.3 shows the required steps towards constructing the prediction model. Different steps involved in performing the prediction are discussed in the rest of this section.

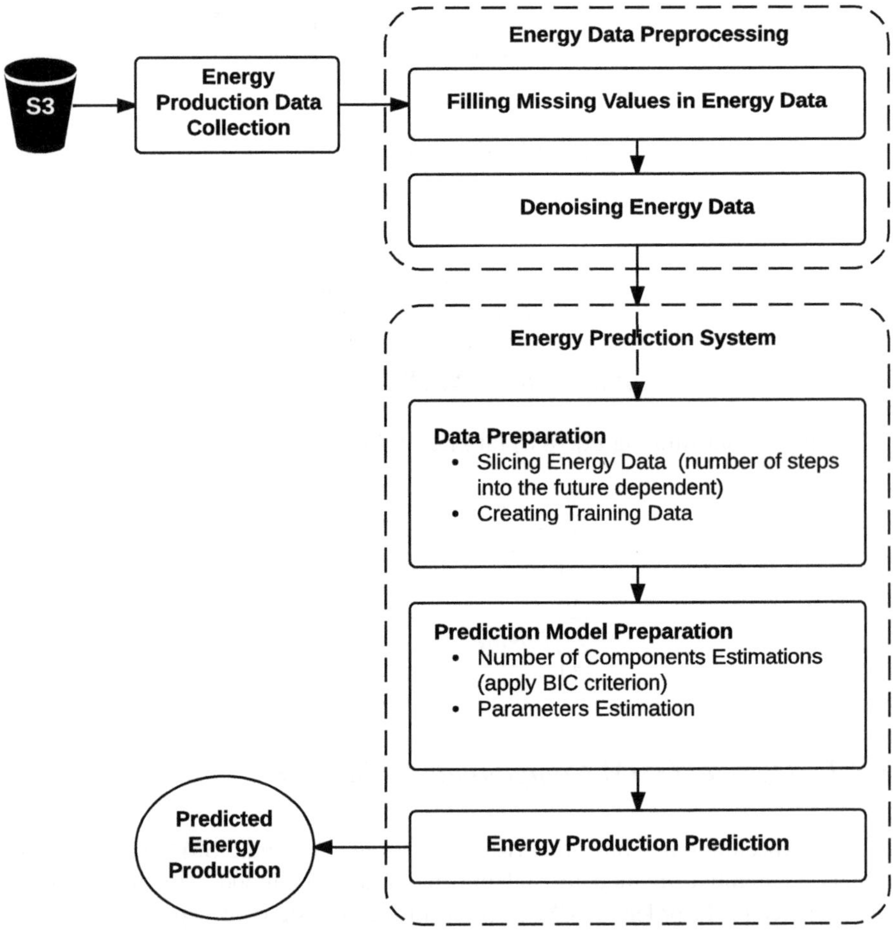

Fig. 8.3 Renewable energy production prediction model

8.4.1 Filling Missing Values in Renewable Energy History Data

Access to accurate history data is critical for prediction. Since having access to perfect history data is not always the case, often there are missing points in time regarding collected history data. Keeping the time-stamp related to each renewable energy data is important to feed into the prediction model. Filling the gaps by simply shifting the energy history data back in time changes the energy data-time mapping. Therefore, we need to fill up the missing values in the collected energy data while keeping each renewable energy's time-stamp. For each collected solar and wind energy, if there are missing data points in the beginning or at the end of a time period, we replicate the first or last observed energy data, respectively. Otherwise, if there are missing energy data in the middle of the time series, we use linear interpolation between the first and the last observed energy data. As presented in Fig. 8.3, filling missing values in the renewable energy history data is part of the preprocessing step, before performing the prediction.

8.4.2 Denoising the Renewable Energy Data

Before training the data and performing the prediction, we need to smooth the collected renewable energy data and remove the sharp acceleration and deceleration of the energy data to achieve a fair prediction. To smooth the history data, we use the fast fourier transform (FFT) algorithm [24] to remove the high frequencies in the energy data and reconstruct it again with only low frequency information.

8.4.3 Training History Data

As shown in Fig. 8.3, we need to prepare the history data to feed into the prediction model. Training set will be constructed according to the following pattern:

$$\mathbf{Z} = \begin{bmatrix} x_1 & x_2 & x_3 & \dots & x_N & y_1 \\ x_2 & x_3 & x_4 & \dots & x_{N+1} & y_2 \\ \vdots & \vdots & \vdots & \ddots & \vdots & \vdots \\ x_T & x_{T+1} & x_{T+2} & \dots & x_{T+N} & y_T \end{bmatrix}, \tag{8.11}$$

where, in our model, $y_i = x_{N+i}$ for $i \in [1, T]$.

To perform the renewable energy production prediction $\hat{y}_{T+1}$, we use the previously observed production values. The granularity of the data history should be equal to the length of the prediction being performed from the last observed renewable energy upto 1-step ahead in time. We denote the granularity of the data history as g, which should be equal to performing the prediction for 1-step ahead into the future (g = 1-step ahead prediction length).

8.4.4 Feature Set Selection

Performing renewable energy prediction requires access to the history data and training the data to estimate prediction model parameters. As stated earlier, we use N previously observed states to predict the next energy production. GMM parameters estimation are driven from running *gmm.fit* on the training set $\mathbf{Z}$ containing history data. The training set is constructed from multiple rows, each equal to a $\mathbf{z} = [\mathbf{x}, x_{t+1}]$ vector, where $\mathbf{x} = [x_{t-N+1}, \ldots x_{t-1}, x_t]$. The elements of $\mathbf{z}$ do not necessarily need to be consecutive observed values. Vector $\mathbf{z}$ elements selection have a major effect on the estimation of the prediction model parameters and accordingly the predicted value of the energy production.

8.5 Prediction Approach and Methodologies

As we mentioned earlier, training the data and filling the training matrix with the right feature set is important to lead us to an accurate prediction. Depending on the time-step ahead into the future that the prediction is taking place, we consider two different approaches to train the data history. To perform the energy prediction for the sth-step ahead into the future, the following two approaches are considered:

- Short-term approach: Selecting every subsequent sth item in the data history.
- Long-term approach: Selecting every subsequent hour:minute corresponding to the hour:minute of the sth-step in the data history.

Moreover, in order to construct the training matrix we consider two different methodologies, as

- Direct multi-step ahead prediction: Direct multi-step ahead prediction (DMSA) performs the energy prediction for s-steps ahead into the future using only the history data. In this approach, the energy production prediction for $\hat{y}_{t+s}$ is independent of the prediction results for energy production before time $t + s$ and is made directly using the data available upto time t.
- Propagated multi-step ahead prediction: Propagated multi-step ahead (PMSA) prediction uses the predicted energy production as an input to the model for next energy production prediction. PMSA uses the $\hat{y}_{t+s-1}$ value as an input to predict the value of $\hat{y}_{t+s}$. The main aim of propagated prediction is to use the results of the previous successful predictions for the next predictions, since prediction results are more accurate for time-steps closer to the last observed energy data.

8.6 Prediction Model Evaluation

This section discusses the experiment setup and the validation of the prediction model. However, as it has been stated before, the accuracy of the prediction model has been tested using the workload demand collected from AWS biggest region, US East, Virginia. We used 1 month of data with granularity 15 min as history data to train the model and predict 7 days ahead into the future. However, due to the confidentiality of the used data set and also our goal to validate the model for renewable energy production, we run a separate set of experiments based on renewable energy production prediction.

8.6.1 *Experiment Setup*

8.6.1.1 Renewable Energy Traces

We use the renewable energy measurements from NREL [21] to calculate solar and wind energy production for a data center. The solar and wind energy traces used in this chapter are the same as the renewable energy used in our previous work [19]. The measurements are with 1 min granularity from May, 1st to May, 29th 2013. We use Global Horizontal Irradiance (GHI) measurements to calculate the output of the solar photovoltaic (PV). The GHI measurements are for PV flat panels on tilted surface at a 45° angle and PV efficiency of 30%. We calculate the solar output based on [25] and the total area for the flat plates is considered to be 100 m^2, derived from the configuration by Solarbayer [26].

To calculate wind energy production, we use the proposed model by Fripp et al. [27]. We feed the wind speed, air temperature, and air pressure, derived from NREL measurements, to the model to calculate wind power at the data center, assuming the data center uses a GE 1.5 MW wind turbine.

8.6.1.2 Benchmark Prediction Models

We compare the results of the prediction model against three different models. Naive that assumes prediction at each point in time is the same as the previously observed value, $\hat{y_{t+1}} = y_t$, linear regression [28] and random forest [29].

8.6.2 *Prediction Analysis Metrics*

We investigate the performance of the prediction model by studying the following quality metrics:

8.6.2.1 Bounded Predicted Values

We use bounded predicted values as a measure to quantify the percentage of the predicted values around $x\%$ of the actual values. This is a good measurement to know for different prediction models, what is the percentage of the predicted values bounded within an error margin (e.g., $\pm 20\%$).

8.6.2.2 R-Squared

In analyzing the accuracy of a prediction, a good prediction model would have the predicted versus actual values as close to the 45° line, as shown in Fig. 8.5. R-squared is a statistical measure that shows how close the predicted values are to the actual values. R^2 gives an intuitive measure of the proportion of the predicted values that could be explained by the actual values. In other words, an R^2 with value x means that $x\%$ of the prediction variation is explained by the actual values.

R^2 value is between 0 and 100%. The higher the R-squared, the better the prediction fits the actual values. If a prediction model could explain 100% of the variance, the predicted values would always equal the actual values and therefore, all the data points would fall on the 45° line.

8.6.2.3 Standard Error

Standard error (S), same as R^2, tells us how well the predicted and actual values would fall on the same line. Standard error is the average distance between the predicted and the actual values. The smaller the S the better the prediction and indicates that the predicted and actual values fall on the 45° line. Moreover, standard error is a good indication to show the accuracy of the prediction. A standard error with value s tells that approximately 95% of the predicted versus actual values fall within $\pm 2 \times s$ of the 45° line.

8.6.2.4 Mean Absolute Error (MAE)

Using a metric that measures the average magnitude of the errors is always useful and indicates how big of an error can be expected from the prediction on average. A perfect prediction would have a *MAE* zero. Since *MAE* is skewed in favor of large errors (prediction outliers), we need to use other metrics such as pth-percentile to better validate the accuracy of the prediction model.

8.6.2.5 P-Percentile

P-percentiles are useful to know the distribution of the prediction error. A pth percentile of a distribution shows that roughly $p\%$ of the error values are equal to or less and $(1-p)\%$ of the error values are larger than that number. Percentiles range in $[0, 100]$. The 0th-percentile shows the min and 100th-percentile shows the max value in a distribution. We measure the pth percentiles on the absolute values of the prediction error ($|\hat{y}-y|$). This way we focus on the unsigned errors and measure how close the prediction and actual values are together, without considering the direction of the error.

It should be noted that when reporting percentiles, we need to consider that if the data distribution is heavy-tailed (right-skewed), significant outliers could be hidden, even not reflected in 90th or 99th percentiles. Therefore, we also report $p-100$ which shows the maximum error value in the prediction.

8.6.3 Prediction Results and Analysis

In the following, we validate the accuracy of the proposed prediction model using the renewable energy measurements from NREL [21]. From the collected renewable energy levels for May 2013, we consider the first 3 weeks as the data history to train the model and the last 8 days as test data to verify the prediction accuracy. We run the prediction model on the previously observed renewable energy production (the data history) to predict the renewable energy level for the next 15 min with the granularity of 1 min. Then, we move the data history window 15-min ahead to predict the next 15 min. We repeat this till we predict 8 days of renewable energy level.

Since the prediction window size is relatively small, 15 min, we use the short-term approach, discussed in Sect. 8.5, to fill the elements in the training matrix. Moreover, we use DMSA methodology, which is independent from the newly predicted values, for feature set selection and training the data. We defer applying long-term approach and PMSA methodology for the interested reader. However, in the carried experiments using AWS data to perform 1-week ahead prediction, we applied short-term approach for predictions up to 36 h ahead in time and long-term approach beyond that time.

Figure 8.4 shows the renewable energy production prediction against the actual values for 8 days. We also demonstrate the GMM prediction results using scatter plot, Fig. 8.5. As it can be seen in this figure, we measure the percentage of the predicted values bounded within ±10% and ±20% relative error and each with considering an absolute error of 5 and 10 kWh, respectively. Using an absolute error constraint on top of the relative error margins prevents small errors to affect our decision making. This has been stated as bounded predicted values in Table 8.1.

We use the previously discussed prediction analysis metrics to evaluate the accuracy of the prediction. The results are presented in Table 8.1. The prediction model column states GMM model and other benchmark models used in our analysis. The results show that almost all the predicted values in GMM (≈ 100%) fall within

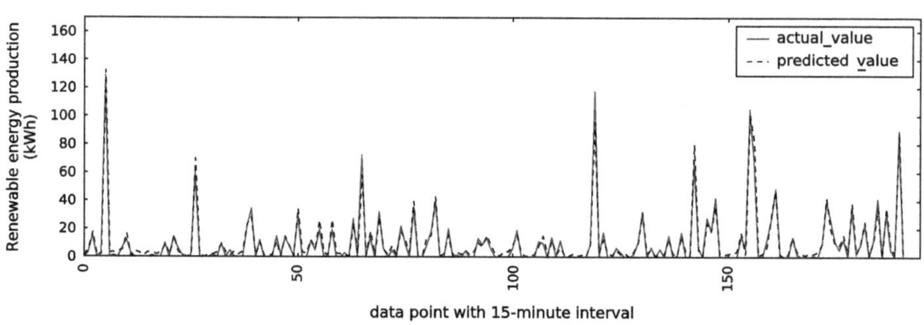

Fig. 8.4 Results of prediction model for 8 days period of renewable energy production for 15-min ahead prediction

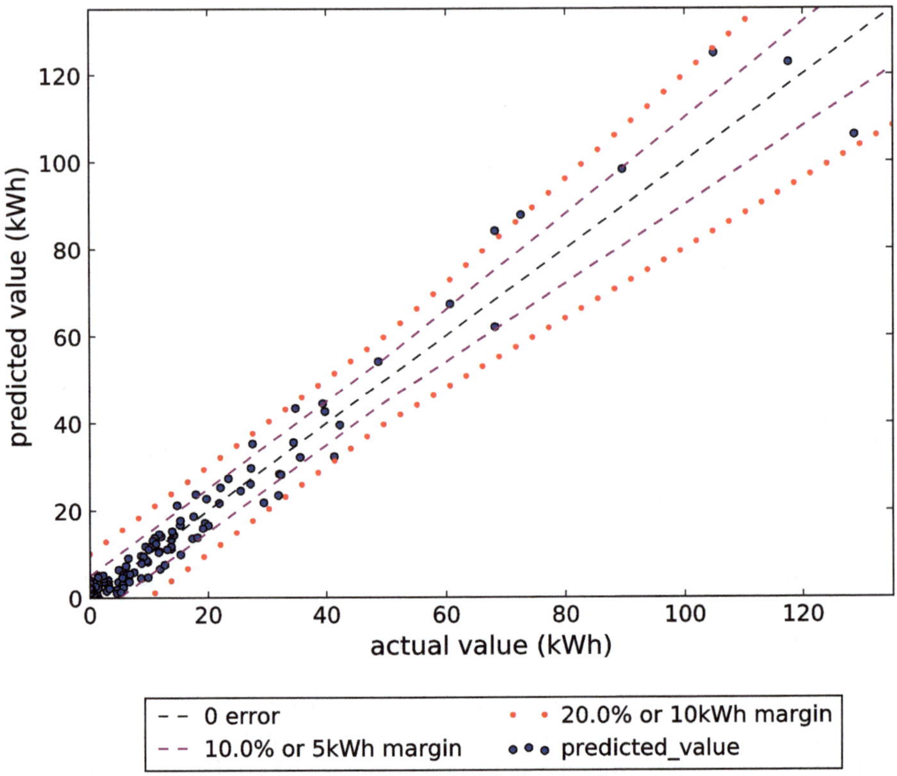

Fig. 8.5 Predicted vs. actual values for 8 days period of renewable energy production for 15-min ahead prediction with ±10% and ±20% around the actual value

Table 8.1 Prediction accuracy under different quality metrics

Prediction model	Bounded predictions (%)	R^2 (%)	*S*	*MAE*	*P*-90	*P*-99	*P*-100
GMM	99.48	97	0.18	2.42	4.16	4.39	21.76
Linear regression	89.81	86.34	0.21	3.97	6.78	8.01	25.72
Random forest	81.27	77.71	0.43	5.45	9.63	11.84	31.65
Naive	47.41	0.01	1.7	16.04	35.56	118.34	128.67

20% of observed actual values, whilst linear regression, random forest, and naive model all are lower than GMM. Even checking bounded predictions bounded within ±10% of the actual values is still close to perfect prediction (97.39%). This means GMM can predict renewable energy with considerably high precision almost similar to real time measuring.

In the rest of the reported metrics, R^2, S, MAE, $P-90$, $P-99$, and $P-100$, GMM is performing better than the rest of the models. Having R^2 of 97% shows that almost all the predicted values are aligned and could be explained by the actual values. Moreover, as per the measured MAE, the prediction error on average is 2.42 kWh, which is a negligible value.

8.7 Summary

In this chapter, we presented a short-term renewable energy production prediction to predict the renewable energy level for many time-steps ahead into the future. The proposed model is based on the Gaussian mixture model and uses history data to train itself and predict the next level of renewable energy in a data center. Knowing the future level of renewable energy helps the cloud provider to make an informed decision to migrate the VMs in the absence of the renewable energy in a data center to a data center with excess renewable energy. This way, the cloud provider can maximize the usage of renewable energy.

To validate the accuracy of the proposed model, we used renewable energy measurements by NREL. The prediction results show that GMM model can predict up to 15 min ahead into the future with nearly 98% precision around $\pm 10\%$ of the actual values. This means that cloud provider can perform online VM migrations with performance close to the optimal offline, that has the full knowledge of the future level of renewable energy.

References

1. Buyya R, Yeo CS, Venugopal S, Broberg J, Brandic I. Cloud computing and emerging it platforms: vision, hype, and reality for delivering computing as the 5th utility. Futur Gener Comput Syst. 2009;25(6):599–616.
2. Mell P, Grance T. The NIST definition of cloud computing. Natl Inst Stand Technol. 2009;53(6):50.
3. Google Apps. https://apps.google.com/. Accessed 18 Apr 2017.
4. Salesforce. https://www.salesforce.com/. Accessed 18 Apr 2017.
5. Google App Engine. https://appengine.google.com/. Accessed 18 Apr 2017.
6. Microsoft Azure. https://azure.microsoft.com/. Accessed 18 Apr 2017.
7. Vecchiola C, Chu X, Buyya R. Aneka: a software platform for .NET-based cloud computing. High Speed Large Scale Sci Comput. 2009;18:267–95.
8. Varia J. Best practices in architecting cloud applications in the AWS cloud. Cloud computing: principles and paradigms. Hoboken, NJ: Wiley; 2011. P. 457–90.
9. Google Cloud. https://cloud.google.com/. Accessed 18 Apr 2017.
10. Rackspace. https://www.rackspace.com/. Accessed 18 Apr 2017.
11. Mankoff J, Kravets R, Blevis E. Some computer science issues in creating a sustainable world. IEEE Comput. 2008;41(8):102–5.
12. Smith JW, Sommerville I. Green cloud: a literature review of energy-aware computing. Dependable Systems Engineering Group, School of Computer Science, University of St Andrews, UK, 2010.
13. NRDC and Anthesis. Scaling Up Energy Efficiency Across the Data Center Industry: Evaluating Key Drivers and Barriers. 2014.
14. AWS and Sustainable Energy. http://aws.amazon.com/about-aws/sustainable-energy/. Accessed 13 Mar 2017.
15. Facebook Installs Solar Panels at New Data Center. http://www.datacenterknowledge.com/archives/2011/04/16/facebook-installs-solar-panels-at-new-data-center/. Accessed 13 Mar 2017.

16. Facebook in Fort Worth: Our newest data center. https://code.facebook.com/posts/1014459531921764/facebook-in-fort-worth-our-newest-data-center/. Accessed 13 Mar 2017.
17. Renewable energy. http://www.google.com/green/energy/. Accessed 13 Mar 2017.
18. Microsoft To Use Solar Panels in New Data Center. http://www.datacenterknowledge.com/archives/2008/09/24/microsoft-uses-solar-panels-in-new-data-center/. Accessed 13 Mar 2017.
19. Khosravi A, Toosi AN, Buyya R. Online virtual machine migration for renewable energy usage maximization in geographically distributed cloud data centers. Concurrency and Computation: Practice and Experience, 2017. doi:10.1002/cpe.4125.
20. Titterington DM, Smith AFM, Makov UE. Statistical analysis of finite mixture distributions. 1985.
21. Measurement and Instrumentation Data Center (MIDC). http://www.nrel.gov/midc/. Accessed 25 Jan 2017.
22. Dempster AP, Laird NM, Rubin DB. Maximum likelihood from incomplete data via the EM algorithm. J R Stat Soc Ser B Methodol. 1977;39:1–38.
23. Schwarz G, et al. Estimating the dimension of a model. Ann Stat. 1978;6(2):461–4.
24. Cooley JW, Tukey JW. An algorithm for the machine calculation of complex fourier series. Math Comput. 1965;19(90):297–301.
25. Photovoltaic Education Network. http://pveducation.org/. Accessed 25 Jan 2017.
26. Solarbayer. http://www.solarbayer.com/. Accessed 25 Jan 2017.
27. Fripp M, Wiser RH. Effects of temporal wind patterns on the value of wind-generated electricity in California and the Northwest. IEEE Trans Power Syst. 2008;23(2):477–85.
28. Montgomery DC, Peck EA, Vining GG. Introduction to linear regression analysis. New York: Wiley; 2015.
29. Breiman L. Random forests. Mach Learn. 2001;45(1):5–32.

Chapter 9
Optimal Sizing of a Micro-Hydrokinetic Pumped-Hydro-Storage Hybrid System for Different Demand Sectors

S.P. Koko, K. Kusakana, and H.J. Vermaak

9.1 Introduction

Pressure on reduction of greenhouse gases emission level as well as on the predicted end of fossil fuel resources is a global concern. The use of environmentally friendly renewable energy sources (RESs) to supply clean and sustainable alternative energy is a solution to mitigate such concern. However, the use of a single technology-based renewable energy (RE) system leads to high investment costs and low reliability due to the intermittent and uncertain nature of RESs [1]. It does not guarantee a continuous uninterrupted power supply and may lead to oversizing of the system. Hybrid renewable energy system (HRES) is a solution to enhance the reliability of the supply at a low maintenance cost [2]. However, uncertainties associated with RESs make HRES sizing a challenging task [3]. By incorporating energy storage system (ESS) to store excess energy can offset the operational uncertainties of RESs in a HRES. This offers a solution to address the mismatch between the demand and the RE output; and also a possibility to reduce the size of the renewable energy system [4]. Among different ESS, pumped-hydro-storage (PHS) is the most promising technology to increase the RE penetration levels in power systems [5]. Due to its lowest cost of energy per cycle, it has proved to be commercially viable by offering 100% of energy autonomy in rural areas [6, 7]. It offsets the commonly used storage batteries by providing larger number of full charge–discharge cycles.

S.P. Koko (✉) H.J. Vermaak
Department of Mathematics, Science and Technology Education, Central University of Technology, Free State, Bloemfontein, South Africa
e-mail: skoko@cut.ac.za; hvermaak@cut.ac.za

K. Kusakana
Department of Electrical, Electronic and Computer Engineering, Central University of Technology, Free State, Bloemfontein, South Africa
e-mail: kkusakana@cut.ac.za

W. Rivera (ed.), *Sustainable Cloud and Energy Services*,
DOI 10.1007/978-3-319-62238-5_9

When planning to construct a HRES, it is important for a designer to select the optimum system configuration satisfying the primary load. Each component of the HRES needs to be correctly sized to ensure the design of an efficient, reliable, and economic hybrid system. Under-sizing a HRES may often unmatch the energy supply and the load demand whereas over-sizing may result into higher capital costs and inefficient use of the HRES [8]. Several software tools are available for designing and evaluating the performance of the HRES as shown in [9]. Hybrid optimization model for electric renewable (HOMER) software is one of the commonly used tools for optimal sizing of a HRES [1, 10]. It determines the optimal size for off-grid and grid-connected systems and can also generate optimization results for variable inputs in order to enable sensitivity analysis with the aim of finding the best configuration based on the given site conditions (load, resources, and components sizes and costs).

Among different renewable energy technologies, hydrokinetic technology is currently gaining more considerable attention. Studies have exposed the potential benefits of using micro-hydrokinetic (MHK) technology in rural electrification. It has proved to offer a reliable and cost-effective electrification solution when compared to solar and wind in areas with flowing water resource [11, 12]. Moreover, it has proved to offer an optimal performance when combined with PHS system instead of battery storage system [13]. Hence, following the above-mentioned benefits, this study focuses on determining the optimal size of a river-based micro-hydrokinetic pumped-hydro-storage (MHK-PHS) hybrid system.

Most sizing and energy optimization studies have concentrated mainly on wind, solar PV, conventional hydro and diesel generator technologies as revealed in [1]. Very few sizing and optimization studies have been conducted on hydrokinetic hybrid system as shown in Table 9.1 [12–17]. These studies used the Legacy Version of HOMER to determine the optimal size of the hydrokinetic hybrid system. The drawback is that the Legacy Version of HOMER does not have a built-in hydrokinetic turbine module in its library. As a result, the authors adopted the methodology of using a wind turbine module to model a hydrokinetic turbine system. The assumptions stating that the anemometer height and the turbine hub height must be set equally to disable HOMER from scaling the wind speed data were adopted.

The aim of this study is to determine the optimal size of the river-based MHK-PHS hybrid system when supplying different load profiles (residential, commercial, or industrial). The optimal size will be determined using HOMER Pro Version 3.6.1 which has a built-in hydrokinetic module in its library. Therefore, the results obtained using HOMER Pro Version 3.6.1 will be compared to the results obtained using HOMER Legacy Version methodology. The aim is to validate the best economical approach to be considered for sizing of the hydrokinetic hybrid system. It has been proved that different energy demand sectors such as residential, commercial, and industrial have load profiles with different curves [18, 19]. None of the sizing and energy optimization studies has revealed the potential impact of different load demand profiles on optimal sizing and operation of a HRES. Therefore, the second objective of the study is to investigate the impact brought

Table 9.1 Summary of hydrokinetic studies that used HOMER simulation tool

Authors	Technology	Key findings	Methodology	Comments
Kusakana and Vermaak [14]	Hydrokinetic, solar PV, wind, diesel generator (DG)	The results revealed that the off-grid hydrokinetic technology is the best viable supply option to consider as compared to standalone wind, solar PV, and diesel generator for both rural household and BTS loads	HOMER Legacy Version was used to perform the simulations	Wind turbine module was used to represent hydrokinetic turbine
Yakub et al. [15]	Hydrokinetic	The hydrokinetic energy was investigated and its potentiality was analyzed in Bangladesh. The hydrokinetic potential was found to be around 1.16 MW	HOMER Legacy Version was used to perform the simulations	Wind turbine module was used to represent hydrokinetic turbine
Kusakana [13]	PHS, hydrokinetic, batteries	The feasibility analysis results revealed that the combination of hydrokinetic and micro-PHS offers a cost-effective, reliable, and environmentally friendly solution as compared to the use of the battery storage system	HOMER Legacy Version was used to perform the simulations	Wind turbine module was used to represent hydrokinetic turbine
Koko [16]	Hydrokinetic, solar PV, wind, DG	The results revealed that the off-grid hydrokinetic technology is the best viable supply option to consider as compared to standalone wind, solar PV, and diesel generator for both rural household and school cases	HOMER Legacy Version was used to perform the simulations	Wind turbine module was used to represent hydrokinetic turbine
Kusakana [12]	Hydrokinetic, solar PV, wind, DG	The techno-economic results revealed that the best hybrid system is the combination of hydrokinetic and DG	HOMER Legacy Version was used to perform the simulations	Wind turbine module was used to represent hydrokinetic turbine
Kusakana and Vermaak [17]	Hydrokinetic, solar PV, DG	The results revealed that the hydrokinetic-diesel hybrid system offers the lowest net present cost and lowest cost of energy compared to other standalone supply options	HOMER Legacy Version was used to perform the simulations	Wind turbine module was used to represent hydrokinetic turbine

by each load profile on optimal sizing and operation of the proposed MHK-PHS hybrid system. For proper comparative purpose, the three load profiles have been standardized to have the same daily energy consumption.

9.2 Mathematical Model of the Proposed Hybrid System

The power flow and water flow layout of the proposed off-grid MHK-PHS hybrid system is shown in Fig. 9.1. The hybrid system consists of the micro-hydrokinetic river system, PHS plant, and the demand sector (residential, commercial, or industrial). The motor-pump unit is used to elevate water from the lower reservoir (river) into the upper reservoir during storage process. The excess energy from the hydrokinetic river system is the one used to power the motor-pump unit. Therefore, the storage process will take place only when the load demand is less than the power generated by the hydrokinetic system. If the load demands more than the generated capacity, the stored water in the upper reservoir will then be released to flow back into the river through the turbine. Electricity generated by the turbine-generator unit is then used to supplement the unmet load. To ensure a reliable and sustainable energy supply, the storage capacity of the upper reservoir must be adequately sized. Mathematical modelling of each building block is discussed below.

9.2.1 Hydrokinetic System

Hydrokinetic river system extracts the kinetic energy of the flowing water in a river by making use of the swept area of a turbine rotor blade. Its output power depends on the speed of the flowing water. It can be installed in rivers with a minimum velocity of 0.5 m/s and above [20]. Its operation principle is similar to the one of a

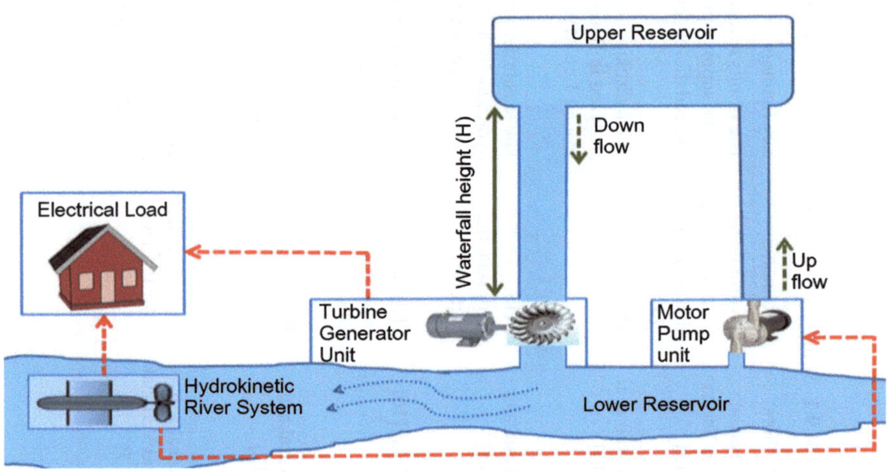

Fig. 9.1 Proposed MHK-PHS hybrid system layout (with electricity flow and water flow)

wind turbine system [21]. However, unlike the wind resource, the main advantage is that the flowing water resource does not suddenly fluctuate within a very short period of time [12]. Since the water density is 800 times greater than the air density, hydrokinetic turbines can extract enough power even at low speed [22–24]. This simply implies that the amount of energy generated by a hydrokinetic turbine is much greater than the one generated by a wind turbine of equal diameter and performance under equal wind and water speed [20]. The energy generated by the hydrokinetic system is expressed using Eq. (9.1) [13].

$$E_{\mathrm{HK}} = 0.5 \times \rho_W \times A \times v^3 \times C_p \times \eta_{\mathrm{HKT\text{-}G}} \times t \tag{9.1}$$

where ρ_W is the water density (1000 Kg/m^3), A is the turbine swept area (m^2), v is the water speed (m/s), C_p is the power coefficient of a turbine performance, $\eta_{\mathrm{HKT\text{-}G}}$ is the overall efficiency of a hydrokinetic turbine-generator unit, and t is the time (s).

9.2.2 Pumped-Hydro-Storage

In this study, micro-PHS plant is used as a means of storing electrical energy during low demand period for later use during energy deficit. The PHS plant used in this study consists of the upper and lower (river) reservoirs as well as separated motor-pump and turbine-generator units. The two reservoirs are linked using two separate penstocks.

9.2.2.1 Motor-Pump Unit

During storage mode certain volume of water needs to be delivered from the lower reservoir into the upper reservoir. Hence, certain power is required by the motor-pump unit to recharge the upper reservoir situated at certain head height. This power is supplied by the hydrokinetic system when the load demands less than the generated power. Hence, during pumping/charging mode the energy used for pumping certain volume of water (m^3) up to a certain height is given by Eq. (9.2) [25]:

$$E_P = \frac{\rho_W \times g \times H \times V}{\eta_P} \tag{9.2}$$

where g is the gravitational acceleration (9.81 m/s^2), H is the water-head height (m), V is the volume of the upper reservoir (m^3), and η_P is the overall efficiency of the pump unit.

9.2.2.2 Upper Reservoir

The volume of the stored water in the upper reservoir must be sufficient to supplement the unmet load demand. The amount of stored energy is proportional to both the volume of the stored water and the waterfall height. The potential energy (joules) of the stored water in the upper reservoir is then expressed using Eq. (9.3) [26].

$$E_S = V \times \rho_W \times g \times H \tag{9.3}$$

When considering the efficiency of the turbine-generator unit, the same upper reservoir's potential energy (kWh) can be expressed using Eq. (9.4) [7, 27].

$$E_S = \frac{V \times \rho_W \times g \times H \times \eta_{\text{T-G}}}{3.6 \times 10^6} \tag{9.4}$$

where $\eta_{\text{T-G}}$ is the overall efficiency of the turbine-generator unit within the PHS system.

9.2.2.3 Turbine-Generator Unit

During energy deficit, the generation process will take place by discharging the upper reservoir. The water flows down through the penstock to enable the turbine-generator units to generate electricity. During the generation process, the volumetric water flow rate (m^3/s) from the upper reservoir through the turbine is directly proportional to the generated power and is determined using Eq. (9.5) [7].

$$Q_{\text{Disch}} = \frac{P_{\text{T-G}}}{\rho \times g \times H \times \eta_{\text{T-G}}} \tag{9.5}$$

where $P_{\text{T-G}}$ is the power generated from the turbine-generator unit.

9.3 Methodology

In this study, HOMER Pro Version 3.6.1 software has been used to determine the optimal size of a proposed hybrid system as submitted to meeting the demand of residential, commercial, and industrial load profiles, respectively. For simulation and optimization analysis, it requires input data such as climatological data, load profile, equipment characteristics, economic and technical data as well as the search space to carry economic optimization of different configurations. HOMER assesses the best configuration of the hybrid system by considering the lowest net present cost (NPC) [28]. The NPC as a financial decision-making technique allows HOMER to analyze the life-cycle cost of the system. The project offering the lowest NPC value reveals the best cost-effective option. HOMER determines the total NPC ($) for each configuration is expressed using Eq. (9.6) [28–30].

$$\text{NPC} = \frac{C_T}{\text{CRF}} \tag{9.6}$$

where C_T is the total annualized costs of the system (US$/yr) and CRF is the capital recovery factor.

CRF is the ratio of the constant annual payments to the present value of receiving the constant annuity over a period of time. It is affected by the selected interest rate and project lifetime and can be determined using Eq. (9.7) [28, 30].

$$\mathrm{CRF} = \frac{i(1+i)^N}{(1+i)^N - 1} \tag{9.7}$$

where N is the number of years to recover the investment and i is the interest rate (%).

The average per unit cost of useful energy as produced by the system configuration to supply the primary load is regarded as a levelized COE. To determine the levelized COE, HOMER considers the quotient of the annualized cost of producing electricity and the total electric load served. Hence, the COE for each configuration is determined using Eq. (9.8) [29, 31].

$$\mathrm{COE} = \frac{C_T}{E_{\mathrm{primary}} + E_{\mathrm{deferrable}} + E_{\mathrm{grid}}} \tag{9.8}$$

where E_{primary} is the primary load served (kWh/y), $E_{\mathrm{deferrable}}$ is the deferrable load served (kWh/y), and E_{grid} is the total grid sales (kWh/y).

During simulations, an interest rate was assumed to be 7% and the project life span was taken to be 25 years. The simulations have been carried out with the intention of meeting the load demand at no capacity shortage. Flowing water resource data of the typical river situated in Kwazulu Natal Province (South Africa), the load profiles and the components costs were fed into HOMER as explained below.

9.3.1 *Load Profiles Description*

Each load profile is modelled separately as a primary load type. The residential load profile of a typical South African residential consumer [32] as well as the commercial and industrial load profiles [19] was used to estimate the daily power curves as shown in Fig. 9.2. For better comparison purpose, the three load profiles were standardized to have the same daily energy consumption of 60 kWh/day. This is denoted by the areas under each curve. The residential load resulted into a peak power demand of 4.42 kW at a small base load of 1.4 kW. The commercial load has a peak power demand of 5.43 kW at a small base load of 0.29 kW while the industrial load has a peak power demand of 7.32 kW at a small base load of 0.14 kW. It can be seen that both the commercial and industrial loads demand most of the energy during the day as compared to the residential load profile. At around 12h00, the industrial load demand drops at a high rate when compared to other working hours. The reason is because unlike the commercial businesses, most industrial businesses allow their employees to simultaneously take a lunch break instead of allowing different break shifts to take place. Hence, this allows the production process to delay a bit.

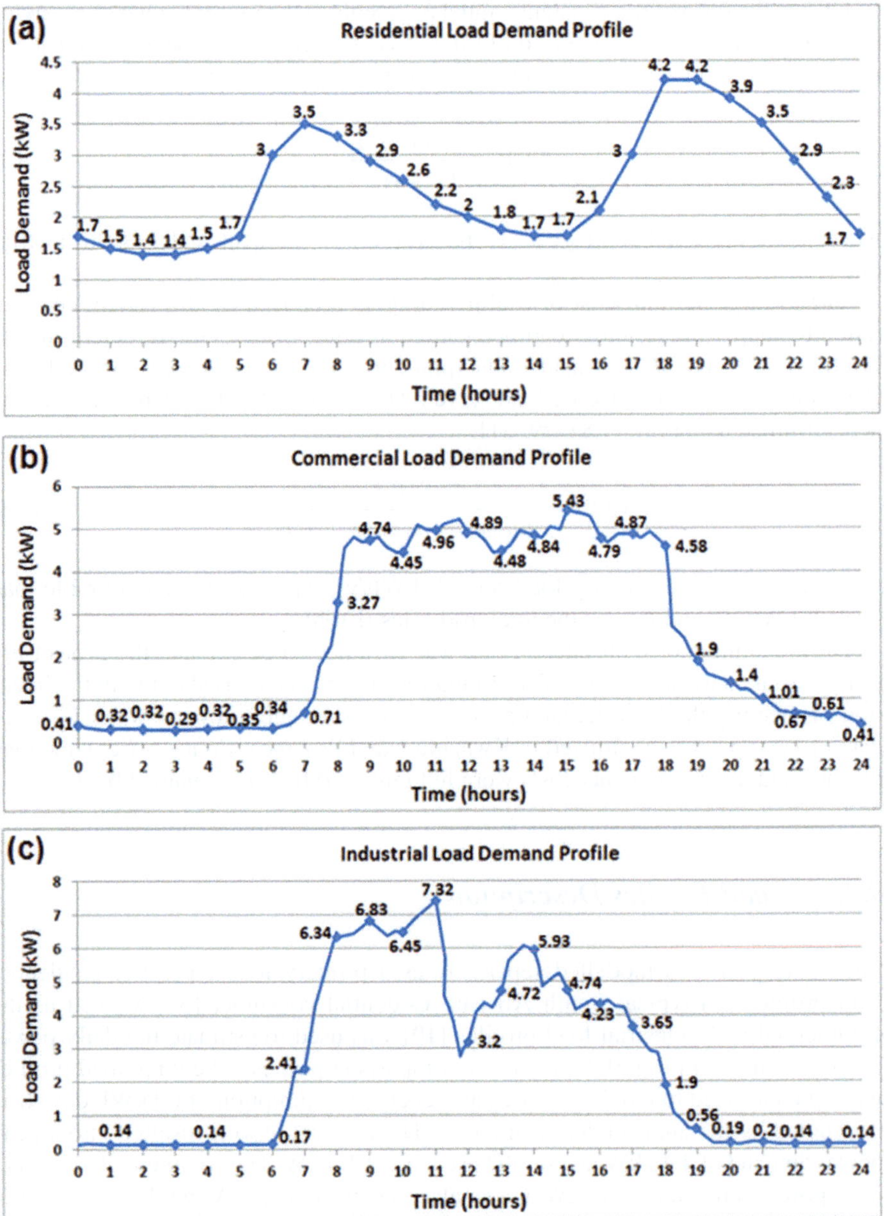

Fig. 9.2 Daily load demand profile for (**a**) residential, (**b**) commercial, (**c**) industrial

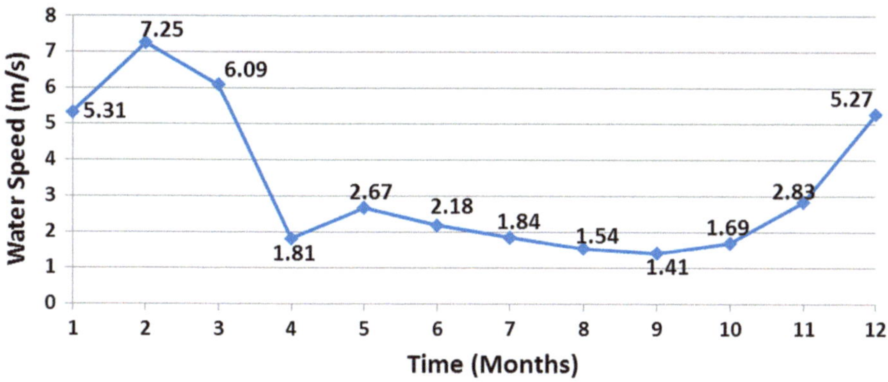

Fig. 9.3 Monthly average water velocity [12–14]

Table 9.2 Components and costs

Components	Capital cost (US$)	O&M cost (US$)	Replacement cost (US$)	Lifetime (years)
1.5 kW hydrokinetic turbine	20,509	355.09	20,509	25
Pumped-hydro storage	3000/kW	180/kW	3000/kW	30

9.3.2 Hydrokinetic Resource Data

Hydrokinetic resource data is necessary to define the flowing water speeds that a hydrokinetic turbine would experience in a typical river. The monthly average water velocity of a typical river in South Africa has been used as input to the hydrokinetic module as illustrated in Fig. 9.3 [12–14]. Although there are some months with insufficient low water speeds, the proposed MHK-PHS hybrid system must be designed to meet the load demand throughout the year.

9.3.3 Components Costs

The performance and costs for each component of the proposed MHK-PHS hybrid system are critical since they are the contributing factors leading to an optimum design configuration. The cost of each component is broken down into capital, replacement, and operation and maintenance (O&M) costs. In this study, all components are assumed to have the replacement costs being equal to the capital cost. Table 9.2 illustrates the sizes and costs of proposed hybrid system's components.

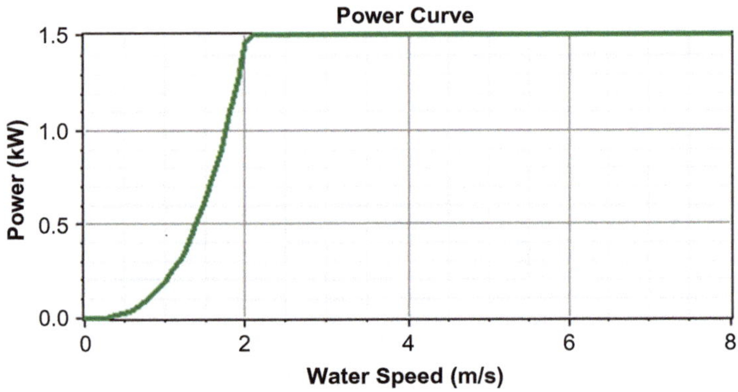

Fig. 9.4 Darius hydrokinetic (DHK) turbine power curve [16]

9.3.3.1 Hydrokinetic System Costs

River-based hydrokinetic turbines are available in a range of 1–10 kW [20]. In this study a generic 1.5 kW direct current (DC) Darrieus hydrokinetic turbine with the swept area of 1.56 m^2 has been selected [16]. It can generate its rated output power at water-flow speed of 2 m/s or above. One unit requires a capital cost of US$15,000 with operation and maintenance (O&M) costs considered to be US$300 per year. Similar to wind turbines, its life span is estimated to be 25 years [33]. The output power curve of the turbine is shown in Fig. 9.4. When the water speed exceeds 2 m/s it is assumed that the output power remains constant at 1.5 kW. An 8 kW, 50 Hz, 230 Vac Victron inverter has been considered in this study to convert DC output power of a hydrokinetic generator into AC power. The cost price of this converter is US$5509 with the O&M cost assumed to be US$55.09 [16]. Since the battery is the only component to be connected to the DC bus when modelling a PHS system [27], the inverter costs will be added to the costs of the hydrokinetic turbine to represent the overall cost of the hydrokinetic system as shown in Table 9.2. The aim is to enable the hydrokinetic turbine to be directly connected to the AC bus.

9.3.3.2 Pumped-Hydro-Storage System Costs

A battery is generally used to model a PHS plant with HOMER software. A battery which is equivalent to an upper reservoir is created and used to model PHS plant [13, 27, 34, 35]. The round-trip efficiency and minimum state of charge (SOC) of the equivalent battery is set to 100% and 0%, respectively. Additionally, HOMER suggests that only the equivalent battery must be connected to the DC bus in order to allow better observation of a PHS performance [27]. To overcome this constraint, a cost-free converter with a round-trip efficiency of 100% must be used to connect the battery to the DC bus.

To represent the upper reservoir, a Trojan T-105 battery was selected and modified to represent the minimum reference volume of the water storage tank. The voltage and the nominal capacity of the Trojan T-105 battery were set to 230 V and 26 Ah, respectively, to allow each battery to store up to a maximum of 6 kWh when providing 1 kW for a maximum of 6 h of discharge. The power delivered by the battery increases with an increase in discharge current [27]. Hence, a maximum discharge current was set to 4.35 A which is equivalent to 0.0926 m^3/s volumetric discharge rate for a minimum volume of 200 m^3 at a head height and round-trip efficiency taken to be 18.35 m and 60%, respectively. The costs of the PHS have been entered into the battery model with a life span taken to be 30 years. The PHS system's installation cost/kW varies between \$2000 and \$4000 [13]. In this study, the installation cost of US\$3000/kW was used during simulations. The O&M costs were assumed to be 6% of the initial capital cost [34].

After determining the optimal number of storage batteries needed for each load demand profile, the total stored energy (kWh) in all batteries will be determined by multiplying the number of batteries by nominal capacity of each battery. HOMER suggests that the energy stored in the batteries is equivalent to the potential energy stored in the upper reservoir. Hence, the total storage capacity in the upper reservoir is the optimum number of batteries multiplied by the storage capacity of each battery and this can be expressed using Eq. (9.9).

$$E_S = n \times C_{\mathrm{Bat}} \times V_{\mathrm{DC}} \tag{9.9}$$

where n is the optimum number of batteries, C_{Bat} is the nominal capacity of one battery (Ah), and V_{DC} is the selected battery voltage.

9.4 Simulation Results and Discussions

The optimal sizing results of the proposed MHK-PHS hybrid system have been obtained using HOMER Legacy Version methodology and using HOMER Pro Version 3.6.1 methodology for comparison reason. When using HOMER Legacy Version methodology, the wind turbine module was used to model/represent the hydrokinetic turbine. As a result, the anemometer height and the turbine hub height were set equally to disable HOMER from scaling the wind speed data. When using HOMER Pro Version 3.6.1 methodology, the actual built-in hydrokinetic module was used to model the hydrokinetic turbine. Three different load profiles (residential, commercial, and residential) were supplied separately as primary load types at 0% capacity shortage. Each load profile was standardized to have a daily energy consumption of 60 kWh for better comparison purpose. This made it possible to analyze the potential impact brought by each load profile on sizing and operation of the proposed hybrid system. Table 9.3 illustrates the optimal configuration results obtained using HOMER Legacy Version methodology and using HOMER Pro Version 3.6.1 methodology. It is important to note that the battery was used to model

Table 9.3 HOMER optimal configuration results

	Residential load		Commercial load		Industrial load	
	HOMER Legacy Version[a]	HOMER Pro Version 3.6.1	HOMER Legacy Version[a]	HOMER Pro Version 3.6.1	HOMER Legacy Version[a]	HOMER Pro Version 3.6.1
Optimization results	7.5 kW hydrokinetic turbine system + 29.9 kWh upper reservoir (5 batteries: 130 Ah)	7.5 kW hydrokinetic turbine system + 17.94 kWh upper reservoir (3 batteries: 78 Ah)	7.5 kW hydrokinetic turbine system + 35.88 kWh upper reservoir (6 batteries: 156 Ah)	7.5 kW hydrokinetic turbine system + 23.92 kWh upper reservoir (4 batteries: 104 Ah)	7.5 kW hydrokinetic turbine system + 41.86 kWh upper reservoir (7 batteries: 182 Ah)	7.5 kW hydrokinetic turbine system + 29.9 kWh upper reservoir (5 batteries: 130 Ah)
Capital cost ($)	117,545	111,545	120,545	114,545	123,545	117,545
Operating cost ($/y)	2635	2292	2808	2464	2980	2636
Net present cost ($)	148,258	138,252	153,263	143,257	158,269	148,263
Levelized COE ($/y)	0.581	0.542	0.601	0.562	0.621	0.581
Total energy production (kWh/y)	53,767	54,152	53,767	54,152	53,767	54,152
Storage autonomy (h)	11.97	7.18	14.36	9.58	16.76	11.97
Excess electricity (kWh/y)	31,882	32,268	31,885	32,271	31,882	32,267

(continued)

Table 9.3 (continued)

	Residential load		Commercial load		Industrial load	
	HOMER Legacy Version[a]	HOMER Pro Version 3.6.1	HOMER Legacy Version[a]	HOMER Pro Version 3.6.1	HOMER Legacy Version[a]	HOMER Pro Version 3.6.1
Turbine-generator hours of operation (h/y)	430	519	613	919	838	1101
Peak turbine-generator POUT (kW)	2.39	2.55	4	2.83	4.51	4.72
Motor-pump hours of operation (h/y)	2526	1693	1104	1299	2532	2588
Peak motor-pump POUT (kW)	2.73	3	4.3	4.15	5.1	4.33
Storage throughput (kWh/y)	369.03	536.81	908.1	1219.7	1057.8	2055.4

[a]Using wind turbine module as a substitute for hydrokinetic turbine

the PHS system. Hence, the rectifier and the inverter represent the motor-pump unit and turbine-generator unit, respectively. The charging of the battery is related to the pumping mode since the upper reservoir is refilled, whereas the discharging of the battery relates to the generation mode since the upper reservoir discharges to generate electricity for supplying the unmet load demand.

9.4.1 Results Comparison of the Two Hydrokinetic Simulation Methods

The optimization results of the two methodologies are shown in Table 9.3. The results show that both methodologies yielded the same hydrokinetic turbine size of 7.5 kW to satisfactorily meet the demand of each load profile. The discrepancy is that the HOMER Legacy Version methodology oversized the storage system by an additional 11.96 kWh (52 Ah) storage capacity for each load profile leading to higher storage autonomy. Despite the high storage capacity computed by the HOMER Legacy Version methodology, HOMER Pro Version 3.6.1 proved that the proposed hybrid system can offer higher excess energy at lower storage capacity. This is a huge saving that will result into lower capital cost. Henceforth, the oversizing of the storage system as computed using HOMER Legacy Version methodology resulted into higher capital cost, NPC, COE, and operating cost when compared to HOMER Pro Version 3.6.1.

9.4.2 Impact of the Load Profiles on Sizing and Operation of MHK-PHS Hybrid System

To compare the impact of each load profile, the optimum configuration results computed by HOMER Pro Version 3.6.1 were used since it has a built-in hydrokinetic module. The optimum configuration results for each load profile proved to require the same hydrokinetic turbine size (7.5 kW) but different storage capacity. As a result, the hydrokinetic system will generate the same amount of power for each load profile throughout the year as shown in Fig. 9.5. It can be seen that when the monthly average water speed is 2 m/s or above, the maximum power production reaches 7.5 kW. This happens during the month of January, February, March, May, June, November, and December.

9.4.2.1 Residential Load Type: Case 1

For supplying the residential load profile, the optimal configuration of the MHK-PHS hybrid system consists of the 7.5 kW hydrokinetic turbine size and a 17.94 kWh storage capacity as calculated by HOMER Pro Version 3.6.1. The aim

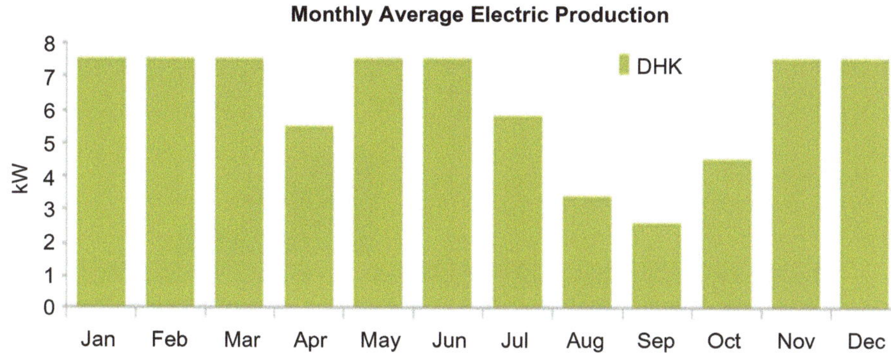

Fig. 9.5 Monthly generated output power from a hydrokinetic system (for each load profile)

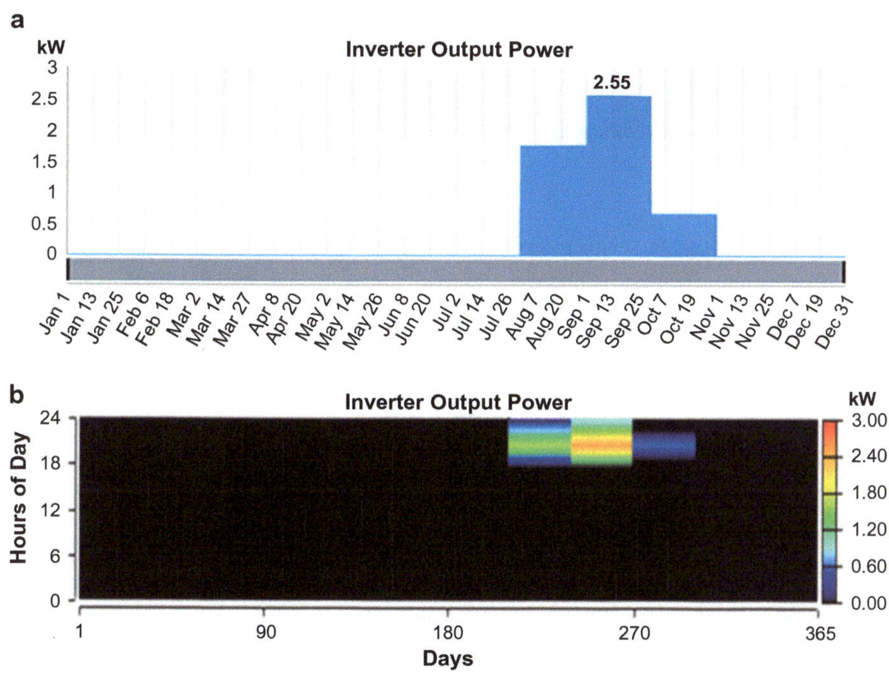

Fig. 9.6 PHS turbine-generator output power (**a**) maximum power (**b**) hourly power

is to ensure 0% of the unmet residential load demand. Hence, the needed volume of the upper reservoir at a water-head of 18.35 m is 600 m^3 when considering the round-trip efficiency of 60%.

Figures 9.6 and 9.7 show the output power of the turbine-generator unit and motor-pump unit, respectively. The turbine-generator output power indicates that the upper reservoir discharges to generate electrical power to be supplied to the unmet

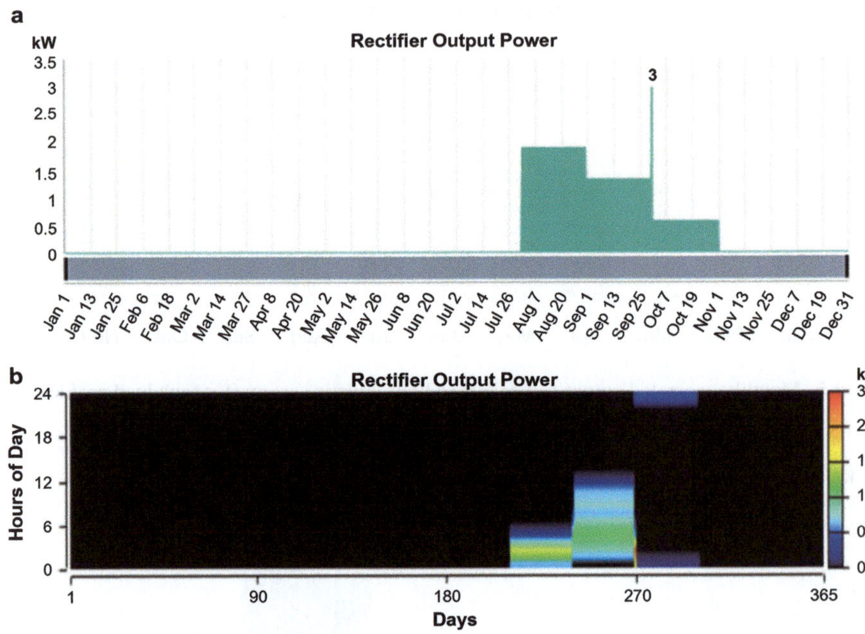

Fig. 9.7 PHS motor-pump output power (**a**) maximum power (**b**) hourly power

load demand. The motor-pump output power indicates that the upper reservoir is recharged (refilled) to store water using excess energy from the hydrokinetic system. Based on Figs. 9.6 and 9.8, it can be seen that the upper reservoir discharges to supply the unmet load demand only during August, September, and October due to the insufficient water speed. The longest discharge hours take place in September between 19h00 and 00h00. This reveals more operational hours for the upper reservoir. As a result, the upper reservoir reaches the lowest SOC of almost 40% of the volume capacity as shown in Fig. 9.8 and starts recharging after 00h00 as shown in Fig. 9.7b. During this month of September, up to a maximum of 2.55 kW output power is demanded by the unmet load from the upper reservoir as shown in Fig. 9.6a. This maximum output power is generally used to determine the size of the required hydro-turbine for the selected flow rate of 0.0926 m^3/s [35]. On the last day of September, the power required to refill the upper reservoir reaches a maximum value of 3 kW as shown in Fig. 9.7a.

9.4.2.2 Commercial Load Type: Case 2

Similar to the residential load, the optimal configuration of the MHK-PHS hybrid system for the commercial load profile consists of a 7.5 kW hydrokinetic turbine. However, the difference is that the commercial load profile requires higher storage

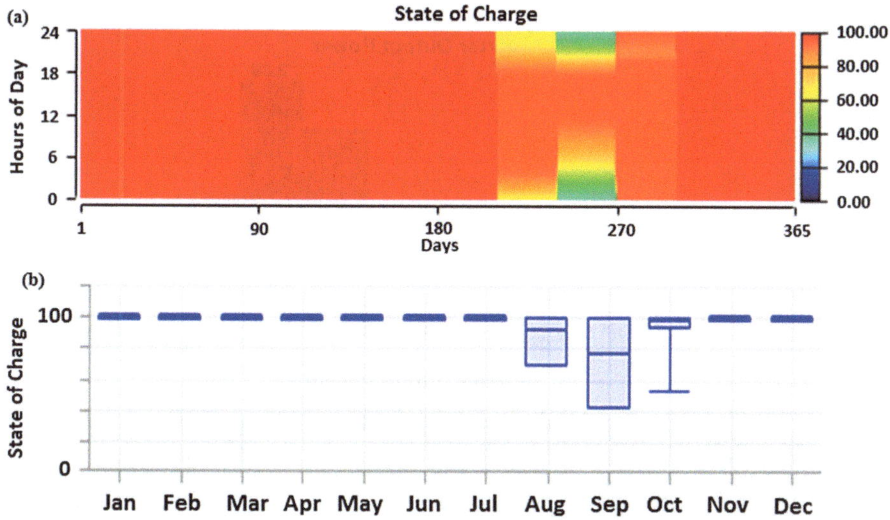

Fig. 9.8 Upper reservoir state of charge (**a**) hourly statistic (**b**) monthly statistic

capacity of 23.92 kWh instead of 17.94 kWh required by the residential load. Therefore, the proposed MHK-PHS hybrid system will require higher capital cost to satisfactorily meet the commercial load demand at no capacity shortage. This resulted into higher NPC, COE, and operating costs. Hence, the needed volume of the upper reservoir at the same water-head of 18.35 m and the same round-trip efficiency of 60% is 800 m^3.

Figures 9.9 and 9.10 show the output power of the turbine-generator unit and motor-pump unit, respectively. Similar to the residential load type, the upper reservoir discharges only in August, September, and October due to the insufficient water speed as shown in Figs. 9.9 and 9.11. The longest discharge duration also takes place in September. However, the difference is that the longest discharge duration takes place between 09h00 and 19h00. This leads to higher operational hours of the upper reservoir as compared to the residential load case. As a result, the upper reservoir approaches the lowest SOC of 4.18 % of the volume capacity as shown in Fig. 9.11 and starts recharging after 19h00 as shown in Fig. 9.10b. During this month, up to a maximum of 2.88 kW output power is demanded by the unmet load from the upper reservoir as shown in Fig. 9.9. Hence, larger hydro-turbine size is required in the PHS system. Fig. 9.10a shows that on the last day of September, the power required to refill the upper reservoir reaches a maximum value of 4.15 kW which is greater than the residential load one.

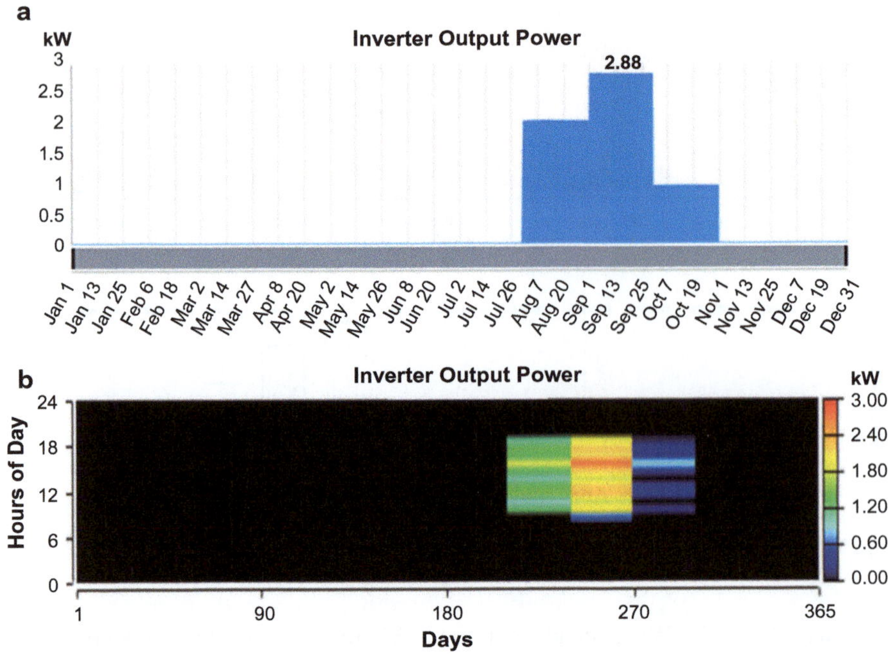

Fig. 9.9 PHS turbine-generator output power (**a**) maximum power (**b**) hourly power

9.4.2.3 Residential Load Type: Case 3

Similar to the residential and commercial load profiles, the optimal configuration of the MHK-PHS hybrid system for the industrial load profile consists of a 7.5 kW hydrokinetic turbine as well. However, the difference is that it requires the largest storage capacity of 29.9 kWh. Therefore, the proposed MHK-PHS hybrid system will require the highest capital cost to satisfactorily meet the industrial load demand at no capacity shortage. This resulted into highest NPC, COE, and operating costs. Hence, the needed volume of the upper reservoir at the same water-head of 18.35 m and the same round-trip efficiency of 60% is 1000 m^3.

Figures 9.12 and 9.13 show the output power of the turbine-generator unit and motor-pump unit, respectively. Unlike the residential and commercial load profiles, the upper reservoir discharges in April, July, August, September, and October as shown in Figs. 9.12 and 9.14. Therefore, this leads to the highest operational hours of the upper reservoir as compared to both residential and commercial load cases. The upper reservoir approaches the lowest SOC of 7.84% of the volume capacity during September especially between 18h00 and 18h59 as shown by the turbine-generator output power in Fig. 9.14. During this month, the unmet industrial load demands up to a maximum of 4.72 kW output power from the upper reservoir as shown in Fig. 9.12. Therefore this leads to the requirement for the largest hydro-

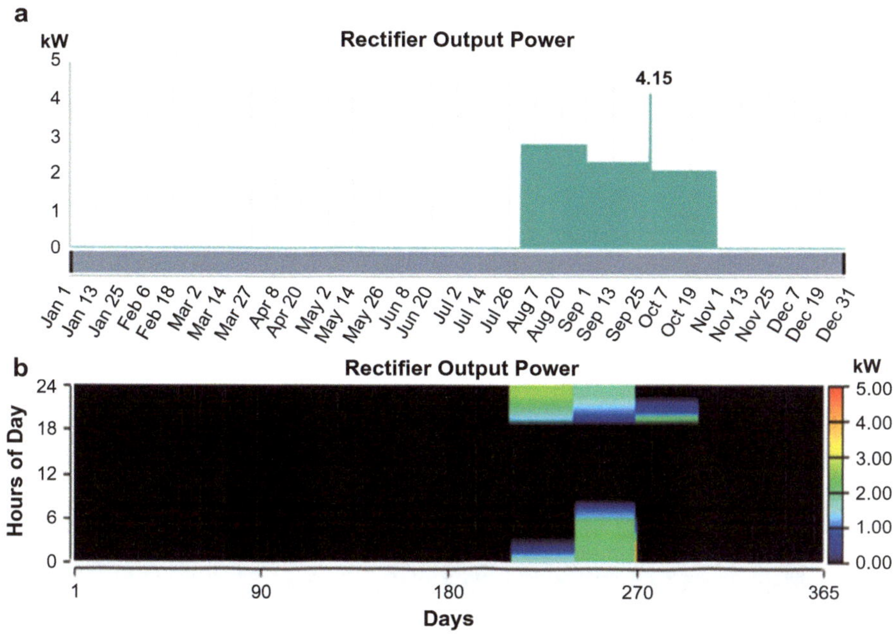

Fig. 9.10 PHS motor-pump output power (**a**) maximum power (**b**) hourly power

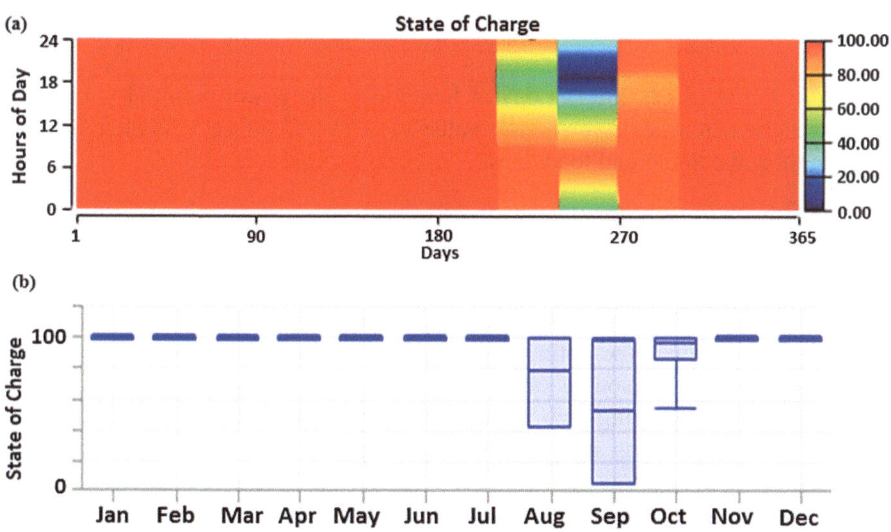

Fig. 9.11 Upper reservoir state of charge (**a**) hourly statistic (**b**) monthly statistic

turbine size as compared to the one required for the residential and commercial load profiles. Industrial load profile allows the motor-pump unit to operate for the longest duration by refilling the upper reservoir from 18h00 till 08h00 in the morning as

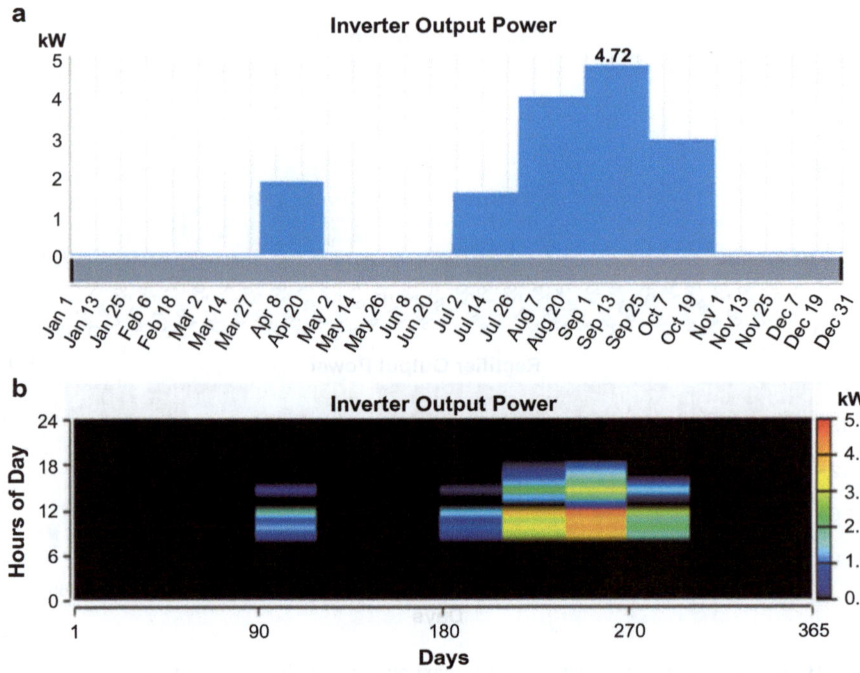

Fig. 9.12 PHS turbine-generator output power (**a**) maximum power (**b**) hourly power

shown in Fig. 9.13. On the last day of September, the power required to refill the upper reservoir reaches a maximum value of 4.33 kW which is greater than the residential and commercial load ones.

9.5 Conclusion

This paper determined the optimal configuration of the MHK-PHS hybrid system using flowing water resource obtained from a typical river of South Africa as a case study by applying two different methodologies. The first methodology was based on modelling a hydrokinetic turbine using wind turbine module as used by previous authors and the second one was based on using a built-in hydrokinetic turbine module available in HOMER Pro Version 3.6.1. Therefore, the objective of comparing the optimal configuration results obtained from these two methodologies will validate the best economical approach. The results showed that the methodology applied in HOMER Legacy Version led to oversizing of the storage capacity by an additional 11.96 kWh. This oversizing constraint resulted into additional system cost leading to higher COE, NPC, and operating costs.

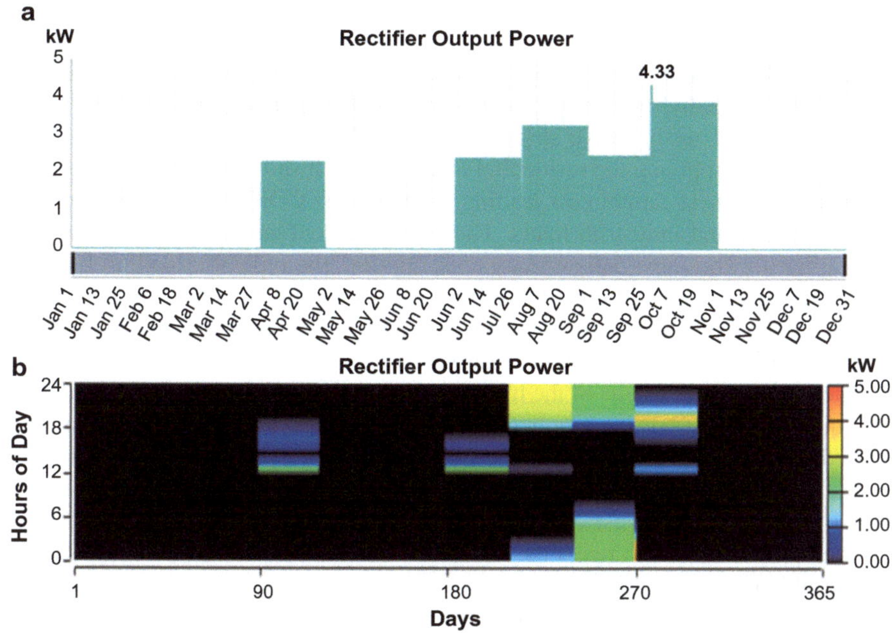

Fig. 9.13 PHS motor-pump output power (**a**) maximum power (**b**) hourly power

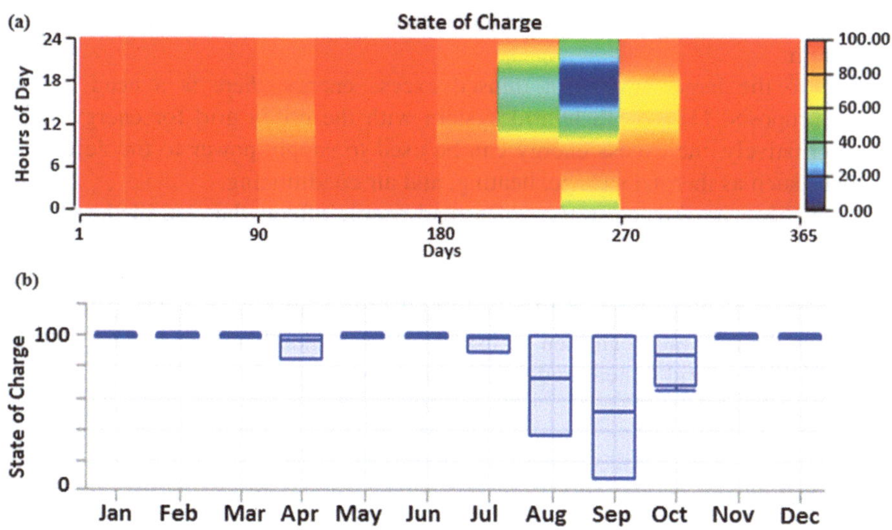

Fig. 9.14 Upper reservoir state of charge (**a**) hourly statistic (**b**) monthly statistic

The second objective of this paper is to analyze the impact brought by different load profiles on optimal sizing and performance of the MHK-PHS hybrid power supply system. Different daily load profiles such as residential, commercial, and industrial having the same daily energy consumption of 60 kWh were considered in this study at 0% capacity shortage. Comparison analysis has been carried out using overall optimum configuration results of both the generation and storage units of the MHK-PHS hybrid system. According to the economic figures of each load profile, HOMER results have shown that the MHK-PHS hybrid system is more superior when supplying the residential load profiles. It shows a good investment by incurring the lowest capital cost leading to the lowest COE and NPC as compared to commercial and industrial loads. The optimum configuration results have also shown that for the same daily energy consumption, the type of a load profile does not affect the size of a hydrokinetic turbine and the amount of annual excess energy. It only affects the required storage capacity leading to different operational hours of charging and discharging the upper reservoir.

The output power results of the turbine-generator unit and motor-pump unit were also investigated. It has been noticed that for the same daily energy consumption, the industrial load profile requires the largest size of a hydro-turbine as compared to the residential and commercial load profiles. The commercial load profile was found to lead to the lowest operational hours of the pumping unit per year. This led the upper reservoir to reach the lowest SOC as compared to the residential and industrial load profiles.

The results of this study have led to the following recommendations:

- To determine the optimal size of a hydrokinetic hybrid system consisting of a storage system, it is advisable to use HOMER Pro Version instead of the Legacy Version to avoid high initial capital costs incurred by the oversized storage system.
- Due to the large amount of unused excess energy, there is a need to link the proposed MHK-PHS hybrid system with the utility grid for energy sales. Alternatively, the excess energy can be used to supply power to the deferrable loads such as thermal storage, heating, and air conditioning.
- It has been proved that a type of a load profile affects the size of the storage system instead of the hydrokinetic turbine size. Therefore, a study needs to be undertaken to determine the daily energy consumption break-even point that will allow the three load profiles to yield the same storage size. The determined energy demand ratio can then be used to know which load will be ideal for the MHK-PHS hybrid system under different daily energy consumptions.

References

1. Bahramara S. Optimal planning of hybrid renewable energy systems using HOMER. Renew Sust Energ Rev. 2016;62:609–20.
2. Fathy A. A reliable methodology based on mine blast optimization algorithm for optimal sizing of hybrid PV-wind-FC system for remote area in Egypt. Renew Energy. 2016;95:367–80.
3. Celik AN. Techno-economic analysis of autonomous PV-wind hybrid energy system using different sizing method. Energy Convers Manag. 2003;44:1951–68.

4. Solomon AA, Kammen DM, Callaway D. Investigating the impact of wind–solar complementarities on energy storage requirement and the corresponding supply reliability criteria. Appl Energy. 2016;168:130–45.
5. Bianchi M, Branchini L, De Pacsale A, Peretto A, Melino F, Orlandini V. Pump hydro storage and gas turbines technologies combined to handle wind variability: optimal hydro solution for an Italian case study. In: Energy procedia, ATI 2015—70th conference of the ATI Engineering Association, vol. 82. 2015. p. 570–576.
6. Zhao H, Wu Q, Hu S, Xu H, Rasmussen CN. Review of energy storage system for wind power integration support. Appl Energy. 2015;137:545–53.
7. Ma T, Yang H, Lu L, Peng J. Technical feasibility study on a standalone hybrid solar-wind system with pumped hydro storage for a remote island in Hong Kong. Renew Energy. 2014;69:7–15.
8. Tazvinga H, Zhu B, Xia X. Energy dispatch strategy for a photovoltaic-wind-diesel-battery hybrid power system. Sol Energy. 2014;108:412–20.
9. Sinha S, Chandel SS. Review of software tools for hybrid renewable energy systems. Renew Sust Energ Rev. 2014;32:192–205.
10. Chang K. A quantile-based simulation optimization model for sizing hybrid renewable energy systems. Simul Model Pract Theory. 2016;66:94–103.
11. Koko SP, Kusakana K, Vermaak HJ. Micro-hydrokinetic river system modelling and analysis as compared to wind system for remote rural electrification. Electr Power Syst Res. 2015;126:38–44.
12. Kusakana K. Techno-economic analysis of off-grid hydrokinetic-based hybrid energy systems for onshore/remote area in South Africa. Energy. 2014;68:947–57.
13. Kusakana K. Feasibility analysis of river off-grid hydrokinetic systems with pumped hydro storage in rural applications. Energy Convers Manag. 2015;96:352–62.
14. Kusakana K, Vermaak HJ. Hydrokinetic power generation for rural electricity supply: case of South Africa. Renew Energy. 2013;55:467–73.
15. Yakub U, Nayeem SM, GolamMostafa SM, Samrat N. An investigation on hydro kinetic energy and analyzing its potentiality in Bangladesh. In: Green energy and technology (ICGET), 2nd international conference, IEEE. 2014. p. 45–49.
16. Koko SP. Techno-economic analysis of an off-grid hydrokinetic river system as a remote rural electrification option [Master's Thesis]. Bloemfontein: Central University of Technology; 2014.
17. Kusakana K, Vermaak HJ. Cost and performance evaluation of hydrokinetic-diesel hybrid systems. In: Energy procedia 2014, the 6th international conference on applied energy—ICAE2014, vol. 61. 2014. p. 2439–2442.
18. Asare-Bediako B, Kling WL, Ribeiro PF. Future residential load profiles: scenario-based analysis of high penetration of heavy loads and distributed generation. Energy Build. 2014;75:228–38.
19. Jardini JA, Tahan CMV, Gouvea MR, Ahn SE, Figueiredo FM. Daily load profiles for residential, commercial and industrial low voltage consumers. IEEE Trans Power Delivery. 2000;15:375–80.
20. Vermaak H, Kusakana K, Koko SP. Status of micro-hydrokinetic river technology in rural applications: a review of literatures. Renew Sustain Energy Rev. 2014;29:625–33.
21. Zhou H. Maximum power point tracking control of hydrokinetic turbine and low-speed high-thrust permanent magnet generator design [Msc Thesis]. Rolla: Missouri University of Science and Technology; 2012.
22. Grabbe M, Yuen K, Goude A, Lalander E, Leijon M. Design of an experimental setup for hydro-kinetic energy conversion. Int J Hydropower Dams. 2009;16(5):112–6.
23. Yuen K, Thomas K, Grabbe M, Deglaire P, Bouquerel M, Österberg D, Leijon M. Matching a permanent magnet synchronous generator to a fixed pitch vertical axis turbine for marine current energy conversion. IEEE J Ocean Eng. 2009;34(1):24–31.
24. Kuschke M, Strunz K. Modeling of tidal energy conversion systems for smart grid operation. In: IEEE power and energy society general meeting, San Diego. 2011. p. 1–3.

25. Jaramillo OA, Rodriuez-Hernandez O, Fuentes-Toledo A. Hybrid wind-hydropower energy systems. In: Stand-alone and hybrid wind energy systems. Cambridge: Woodhead Publishing; 2010. p. 282–322.
26. Akinyele DO, Rayundu RK. Review of energy storage technologies for sustainable power networks. Sustain Energy Technol Assess. 2014;8:74–91.
27. Canales FA, Beluco A. Modelling pumped hydro storage with the micropower optimization model (HOMER). J Renew Sustain Energy. 2014;6:043131.
28. Mamaghani AH, Escandon SAA, Najafi B, Shirazi A, Rinaldi F. Techno-economic feasibility of photovoltaic, wind, diesel and hybrid electrification systems for off-grid rural electrification in Colombia. Renew Energy. 2016;97:293–305.
29. Amutha WM, Rajini V. Cost benefit and technical analysis of rural electrification alternatives in southern India using HOMER. Renew Sust Energ Rev. 2016;62:236–46.
30. Montuori L, Alcázar-Ortega M, Álvarez-Bel C, Domijan A. Integration of renewable energy in microgrids coordinated with demand response resources: economic evaluation of a biomass gasification plant by Homer Simulator. Appl Energy. 2014;132:15–22.
31. Shezan SA, Julai S, Kibria MA, Ullah KR, Saidur R, Chong WT, Akikur RK. Performance analysis of an off-grid wind-PV (photovoltaic)-diesel battery hybrid energy system feasible for remote areas. J Clean Prod. 2016;125:121–32.
32. Eskom. Introduction to Homeflex. 2016. Accessed 09 Dec 2016. Available at: http://www.eskom.co.za/CustomerCare/TariffsAndCharges/Documents/Eskom%20Booklet.pdf.
33. Hiendro A, Kurnianto R, Rajagukguk M, Simanjuntak YM. Techno-economic analysis of photovoltaic/wind hybrid system for onshore/remote are in Indonesia. Energy. 2013;59:652–7.
34. Canales FA, Beluco A, Mendes CAB. A comparative study of a wind hydro hybrid system with water storage capacity: conventional reservoir or pumped storage plant? J Energy Storage. 2015;4:96–105.
35. Iqbal T. Feasibility study of pumped hydro energy storage for Ramea wind-diesel hybrid power system. St. John's: Faculty of Engineering and Applied Science, Memorial University; 2009.

Chapter 10
Impact of Different South African Demand Sectors on Grid-Connected PV Systems' Optimal Energy Dispatch Under Time of Use Tariff

K. Kusakana

10.1 Introduction

The South African electric power system was designed in times when there were adequate energy storage systems, forcing instantaneous consumption. Recent developments in energy storage systems have changed all that, bringing about the introduction of distributed power sources such as solar or wind systems [1].

Solar photovoltaic and wind energy systems have low running costs on condition that they are operated within prescribed constraints, as the power produced tends to displace conventional generation from non-renewable energy sources [2]. However, these renewable energy schemes are not sustainable when operated without proper means of storing energy [3]. Solar irradiation and wind highly variable and they can negatively impact the power generation when operating in a standalone or grid connected mode, causing severe challenges in responding to the demand by the generation [4].

Providing an electricity supply which continuously matches consumer demand is one of the main challenging tasks of any power generation system. Complying with this operational constraint is a major challenge in the distributed power generation industry using renewable sources because the resources may fluctuate in an uncontrolled manner; on the other hand, the fluctuating consumer demand must be satisfied [5].

Based on these, any power generation entity needs to fulfil two exclusive tasks:

- the necessity to maintain a continuous power balance between power production and consumption,

K. Kusakana (✉)
Department of Electrical, Electronic and Computer Engineering, Central University of Technology, Free State, Bloemfontein, South Africa
e-mail: kkusakana@cut.ac.za

W. Rivera (ed.), *Sustainable Cloud and Energy Services*,
DOI 10.1007/978-3-319-62238-5_10

- and the necessity to regulate and manage power flows between the generation, the load and the storage [6].

To adequately meet consumer demands, fast power sources optimized to respond to the needs of the variable demand can be incorporated as part of the supply. Batteries, which can supply the energy stored on demand can be adequately used to solve this problem. They have the unique property to store the excess energy from the generator for further use when load demands are greater than the production. They are also appropriate for minimizing and smoothing out the variability of renewable energy outputs and are being commonly used in these situations [7].

In real time applications, the rapid response to fluctuation load exhibited by battery storage systems is constrained by the size of storage needed to back up unavailability in renewable power sources. This size can be much higher, if not used in conjunction with a conventional generator or is connected to the grid [8, 9].

Dispatch refers to the process by which an electrical grid manages how much power should be flowing from the generating station to the load. It may also include the management of available excess power to or from the battery in the case of a power shortfall, in response to rapid demand changes or to specific load profiles of a demand sector [10]. Industrial and commercial sectors usually have set daily operational schedules making their average load profile quite predictable [11]. Likewise, the power demand from the domestic sector can also be depicted from the user's habits and therefore varies from one location to another [12].

The new scheme from the South African government to promote the use of solar energy systems is favouring the development and implementation of small domestic and other power producers dispersed throughout the local grid. This support from the government usually comes with an obligation for the main electricity supplier to purchase the excess power produced. While the excess power from users supplied to the grid is a source of income to the renewable energy producer, it can also induce several challenges such as the need for adequate bidirectional power connection between the grid and the generating station; and the power generated, consumed or stored at anytime must be meticulously controlled to continuously meet the demand while maximizing the profit by selling the excess to the grid. Battery storage systems enable the power flow in the system to be optimally dispatched or scheduled, allowing the user to take advantage of the difference between prices at their peak through the Time-of-Use tariff (TOU) [13].

Grid-connected PV with battery storage systems is currently gaining considerable attention. Studies have exposed the potential benefits of using this technology in rural electrification. It has proved to offer a reliable and overall cost-effective power when compared to systems without storage or where traditional diesel generators are used as back-up systems [14]. Several research papers have analysed the sizing and scheduling of macro grid-tie PV systems [15–20]. After reviewing these works, it has been noticed that the impact of different load types of users or sectors on the daily operation cost of grid-connected PV system has not been investigated. For this reason, the current paper evaluates the impact of different daily demand sectors in terms of the resulting operation scheduling and corresponding operation costs of the proposed grid-connected photovoltaic system in Bloemfontein. The optimal scheduling of the proposed system has been modelled and simulated using Matlab Simulink. For comparison purposes, the three load profiles, respectively from residential, commercial and industrial sectors, have been depicted and normalized to have the same daily energy consumption levels with different peak demand patterns. The results have shown that for the same daily energy consumption, the type of a load profile affects the grid-connected PV system's operation, resulting in different daily operational costs achieved. Consequently, it can be recommended that in Bloemfontein and South Africa in general.

10.2 Description of the Proposed Microgrid

The model developed will help to determine the optimal scheduling of the system and its corresponding operational costs for the considered operation window, given the variable power required by consumers, the energy from the sun and the fluctuating cost of electricity from the utility grid. Therefore, the initial capital cost is not considered because it is not a variable influencing the operational scheduling of the system. This cost can be considered at a later stage, together with replacement cost, for lifecycle cost analysis, for example, which is beyond the scope of this work.

10.2.1 Schematic Diagram of the System

The photovoltaic generator is used as main supply to consumers. In cases in which there is an excess of energy from the PV generator, the surplus power can either be stored or fed into the grid, depending on the pricing period. For periods in which customer power requirements are higher than the PV generation, the stored energy is released to top up the balance from the demand. This stored energy can also be fed into the grid when the demand is entirely satisfied. Depending on the pricing period, the grid can be used to feed the consumer or to store energy for future use.

From Fig. 10.1, the following control variables can be identified:

- P_1: Power from the PV system for battery charging;
- P_2: Power from the storage used to feed the consumer;
- P_3: Power from the grid for battery charging;
- P_4: Power from the grid for load demand;
- P_5: Power from the PV system for load demand supply;
- P_6: Power from the hybrid system sold to the grid.

These power flows will be the control variables to be optimized.

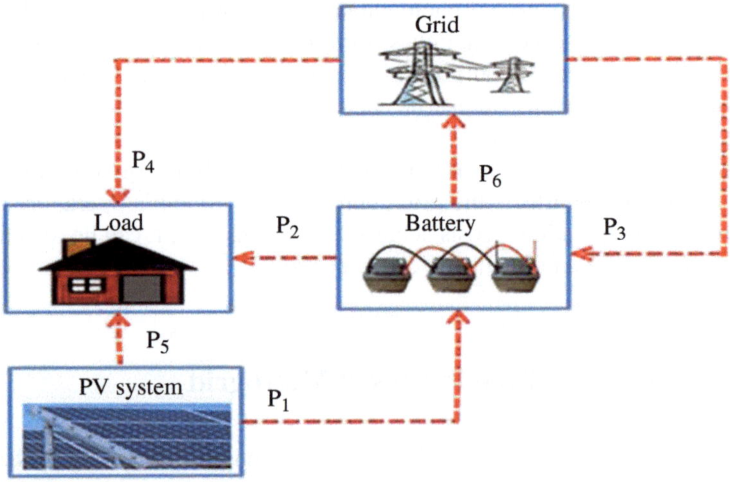

Fig. 10.1 Set-up of the studied microgrid

10.2.2 Photovoltaic Model

A simplified expression of the PV output power can be found in Ref. [21, 22]. For a given size, the output power can be expressed as:

$$P_{PV} = A \times \eta_{PV} \times I \qquad (10.1)$$

With:

- A_{PV} is the surface size of the PV array (m^2);
- η_{PV} is the efficiency PV system;
- I is the solar irradiation (kWh/m^2).

10.2.3 Battery Bank Model

The power balance between the generation and the varying consumer power requirements influences the charge level in the battery (SoC) which can increase, decrease or stay constant. This dynamic can be expressed as:

$$\mathrm{SoC}_{(j+1)} = \mathrm{SoC}_{(j)} + P_{\mathrm{in}(j)}\frac{\Delta t \times \eta_C}{E_{\mathrm{rat}}} - P_{\mathrm{out}(j)}\frac{\Delta t}{E_{\mathrm{rat}} \times \eta_D} \qquad (10.2)$$

where:

- SoC: is the percentage energy level of the storage at any given time;
- η_C: is efficiency linked to the charging process of the battery;
- P_{in}: is the power used to charge the battery (P_1 and P_3);
- P_{out}: is the power used from the battery (P_2 and P_6);

- η_D: is efficiency linked to the discharging process of the storage;
- E_{rat}: is the rated energy or size of the storing system.
- j: is the considered sampling interval ($1 \leq j \leq N$).

Equation (10.2) can be further developed by induction to introduce the initial state of charge as:

$$\mathrm{SoC}_{(j)} = \mathrm{SoC}_{(0)} + \sum_{i=0}^{j-1} P_{\text{in}(j)} \frac{\Delta t \times \eta_C}{E_{\text{rat}}} - \sum_{i=0}^{j-1} P_{\text{out}(j)} \frac{\Delta t}{E_{\text{rat}} \times \eta_D} \tag{10.3}$$

10.2.4 Flowchart of the Proposed Optimization Methodology

The simplified flowchart of the proposed methodology is presented in Fig. 10.2. The main optimization variables are described below:

- Independent variables: Load demand, solar and wind resources, Time-of-Use.
- Dependent variables are all variables which are affected by any change or variation in the input variables. In this case it is manly the battery state of charge (SOC).
- Controlled variable: The powers related to the renewable sources, storage, and the grid are considered as controlled variables.

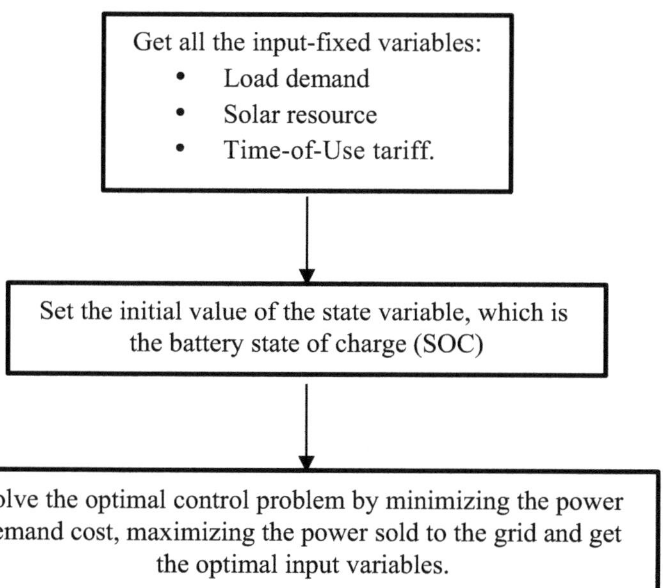

Fig. 10.2 Flowchart of the proposed optimization methodology

10.3 Optimization Model and Proposed Optimal Control Algorithm

The purpose of the model to be developed is to realize minimal running expenses of the proposed microgrid by finding its optimal schedule of operation that will allow a minimum energy obtained from the utility and a maximum injected to the grid given the Time of Use tariff imposed by the grid. More details on the price structure used in this work can be obtained from Ref. [23]: The price of electricity for Bloemfontein in South Africa is presented below:

$$\rho(t) = \begin{cases} \rho_k; t \in T_k, T_k = [7, 10) \cup [18, 20) \\ \rho_0; t \in T_0, T_0 = [0, 6) \cup [22, 24) \\ \rho_s; t \in T_s, T_s = [6, 7) \cup [10, 18) \cup [20, 22) \end{cases} \quad (10.4)$$

With:

- $\rho_k = 2.2225$ R/kWh (electricity rate in peak pricing time interval);
- $\rho_0 = 0.3656$ R/kWh (electricity rate in off-peak pricing time interval);
- $\rho_s = 0.6733$ R/kWh (electricity rate in standard pricing time interval).

10.3.1 *Objective Function*

The objective function is given as Eq. (10.4) where the first part represents the charge of buying power from the utility (to be minimized); the second part is the income realized from feeding power to the utility (to be maximized); the third part is the wearing component cost of whole scheme as explained in Ref. [23].

$$g = \sum_{j=1}^{N} \rho_j \left(P_{3j} + P_{4j}\right) \Delta t - r_k \rho_k \sum_{j=1}^{N} P_{6j} + \sum_{j=1}^{N} a \left(P_{2j} + P_{6j}\right) \Delta t + 24b \quad (10.5)$$

With:

- r_k as a fraction of the rate for the peak pricing time interval;
- ρ_k for selling power during the peak pricing period;
- a is the coefficient of the battery wearing cost and
- b is the hourly wearing cost of the other components.

10.3.2 Constraints

The control variables to be optimized were presented in Sect. 2.1; they have to meet the constraints below:

- PV generator:

As stated in Sect. 2, the PV system can deliver power to consumers and/or feed the storage system. Therefore, at any instance, the power from the PV feeding the demand and/or feed into the storage system must be less than the instantaneous power from the PV generator. This is expressed in the equation below:

$$P_{1(j)} + P_{5(j)} \leq P_{PV(j)}^{\max} \tag{10.6}$$

- Power balance constraint:

The power required to satisfy the consumer must be equal to the linear combination of the powers from the PV, the utility and from the storage system as formulated in the equation below:

$$P_{L(j)} = P_{2(j)} + P_{4(j)} + P_{5(j)} \tag{10.7}$$

- Control variables boundaries

Each control variable can take a set of values between a minimum and a maximum limit for the proposed simulation horizon time as formulated in the equation below:

$$0 \leq P_{1(j)} \leq P_1^{\max} \; (1 \leq j \leq N) \tag{10.8}$$

$$0 \leq P_{2(j)} \leq P_2^{\max} \; (1 \leq j \leq N) \tag{10.9}$$

$$0 \leq P_{3(j)} \leq P_3^{\max} \; (1 \leq j \leq N) \tag{10.10}$$

$$0 \leq P_{4(j)} \leq P_4^{\max} \; (1 \leq j \leq N) \tag{10.11}$$

$$0 \leq P_{5(j)} \leq P_5^{\max} \; (1 \leq j \leq N) \tag{10.12}$$

$$0 \leq P_{6(j)} \leq P_6^{\max} \; (1 \leq j \leq N) \tag{10.13}$$

10.3.3 Optimal Control Method

The different power flows as well as the variable load requirements at any selected sample period *j* influence the charge level in the battery. As for the control variables, the battery SOC can vary between a maximum and a minimum set value. This is given by the expression:

$$SOC^{min} \leq SOC_{(j)} \leq SOC^{max} \tag{10.14}$$

10.3.4 Algorithm Formulation in Matlab

Given the linear nature of the optimization problem developed using the objective function and constrains given from Eq. (10.5) to Eq. (10.9); it can be resolved using the linear programming solver in Matlab with the following syntax [23]:

$$\min g(x), s.t \begin{cases} Ax \leq b \\ A_{eq}x = b_{eq}, \\ l_b \leq x \leq u_b \end{cases} \tag{10.15}$$

With:

- $g(x)$: represents the objective function;
- A_{eq} and b_{eq}: represent the equality constraints parameters;
- A; b: represent the inequality constraints parameters;
- l_b; u_b: represent the inferior and superior limits of the variables.

10.4 Application to Bloemfontein Case (South Africa)

10.4.1 Presentation of the Studied Area

The solar data used for the simulation were collected for the case of Bloemfontein (29° 12′ S 26° 07′ E). This region is characterized by sunny weather which can be adequately used to generate energy using solar systems.

In this area, pyranometers have been installed for the purpose of obtaining solar radiation data. A data acquisition system was used to record the data and average it over 5 min intervals and was this recorded in the memory of the datalogger. To study the performance of the proposed grid connected system, the used data were averaged over each 30 min which is the selected simulation sampling time. These data are available from Ref. [23].

10.4.2 Load Profiles Description

As stated in the introduction, this study focuses on analysing the impact brought by residential, commercial and industrial use on the daily running expenses and schedule of the considered microgrid working under TOU. For this purpose, the impact of different load profiles on the optimal power scheduling of the considered microgrid has been determined using three different load profiles having the same energy consumption of 75 kWh/day.

Figure 10.3, representing a residential load profile, shows that the power demand has a low peak of 5.5 kW in the morning at around 09:00, a variable power demand throughout the daytime between 11:00 and 17:00, and a maximum demand of 8 kW.

Commercial and industrial demand profiles are represented in Figs. 10.4 and 10.5, respectively. They build up in the morning to a to reach a peak of 5.2 kW for commercial and 4.3 kW for industrial, then fall in the evening to reach their minimum during the night-time.

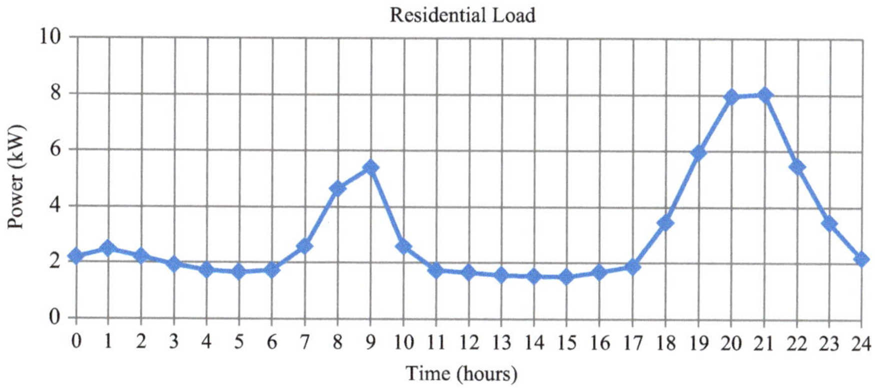

Fig. 10.3 Proposed residential load profile

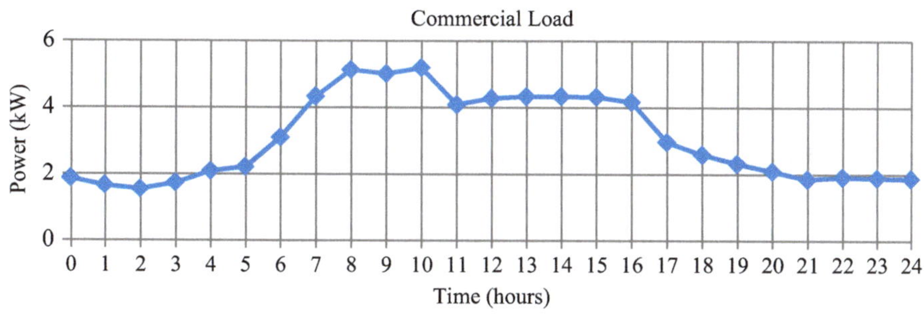

Fig. 10.4 Proposed commercial load profile

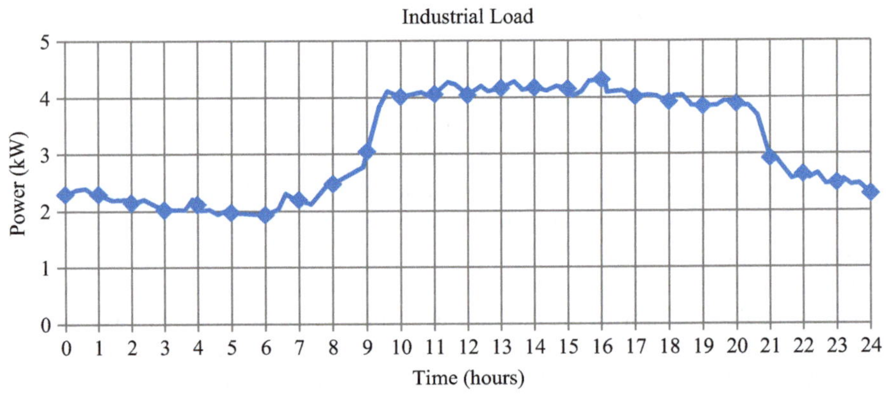

Fig. 10.5 Proposed industrial load profile

10.4.3 The Components Characteristics

The size of the hybrid system's components used in this study can be found in our previous work available in Ref. [23] where all the parameters and specifications necessary for the simulation are explained in detail.

10.5 Discussions

As described in the introduction, the current analysis is based on the daily operational behaviour of the considered microgrid. The performance of the system supplying the three different types of load under TOU will be discussed and categorized to find out which of the demand sectors is more likely to generate more income when the hybrid system supplying its demand is connected to the grid. It is important to highlight that the hybrid system size, architecture, control settings, PV resources as well as energy requirements are the same for the three cases, only the load profiles (power demands) are different.

10.5.1 Residential Demand Supplied by the Proposed System

- System performance during off-peak pricing time interval [00:00, 06:00)

Figure 10.6a illustrates the daily consumer's power demand for the considered winter day in Bloemfontein. During this pricing time interval, the storage system alone is used to respond to the consumer's demand as pointed out in Fig. 10.6c; the photovoltaic system and the utility do feed the consumer as illustrated in Fig. 10.6b, d.

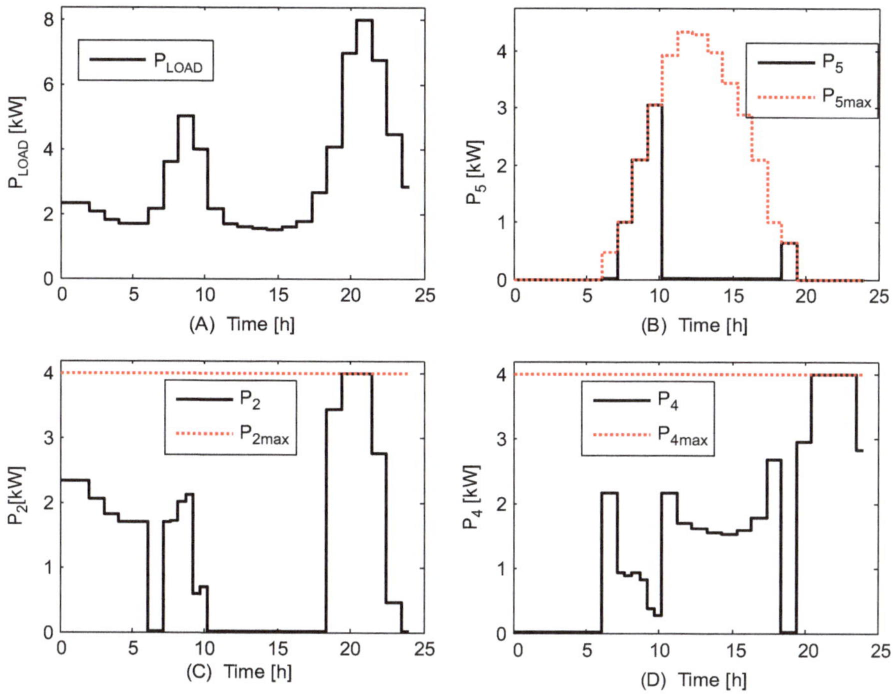

Fig. 10.6 Residential case: Demand side power scheduling

Figure 10.6a shows how the battery is discharged within its operational limits while supplying the load. From Fig. 10.6b, c it can be seen that during this off-peak pricing time interval, neither the utility and the PV supply the battery. Figure 10.6d reveals that there is no power fed into the utility during this off-peak pricing time interval.

- System performance during standard pricing time interval [06:00, 07:00)

All through the first standard pricing time interval taking place between 06:00 and 07:00, the consumer's power requirement is exclusively met by the utility as revealed from Fig. 10.6d. The Photovoltaic systems and battery storage system are not used to feed the consumer, as illustrated in Fig. 10.6b, c.

From Fig. 10.7c it can be noticed that the storage system is being recharged by the utility, this is reflected on Fig. 10.7a where an increase in the storage's SoC is observed. Figure 10.7d reveals that there is no income generated because the utility does not receive any power from the renewable source or from the storage system during this time interval.

- System performance during peak pricing time interval [07:00, 10:00)

During this peak pricing time interval, the consumer is supplied with power from the photovoltaic combined with the storage system (Fig. 10.6b, c) with a small contribution from the utility. The SoC is decreasing because of the power

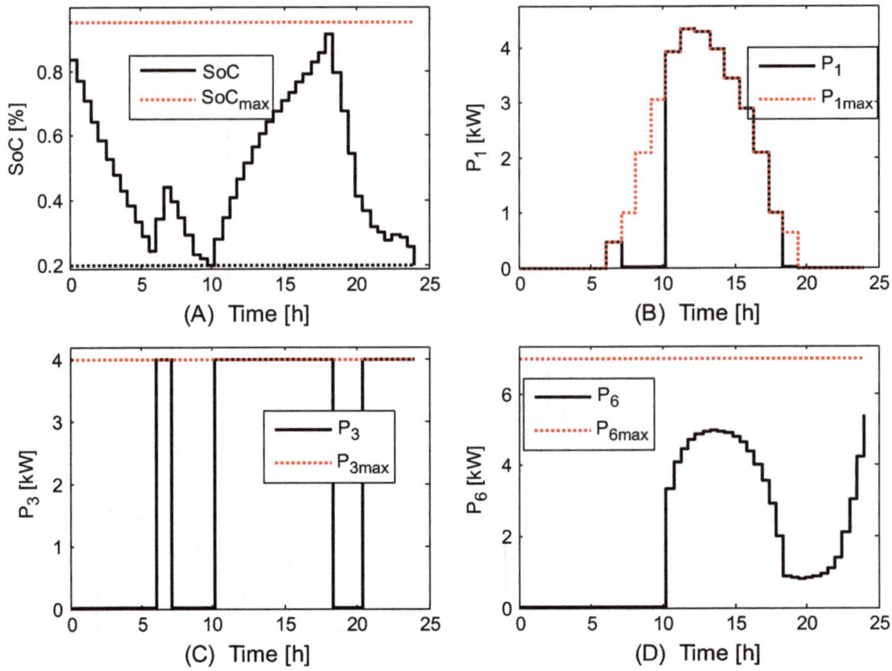

Fig. 10.7 Residential case: Storage side power scheduling

flowing from the storage to the consumer, as shown in Fig. 10.7a. The model has been developed to maximize the power fed into the grid with the aim of generating income at the consumer side. However, it can be seen from Fig. 10.7d that no profit is generated during this peak pricing time interval due to the priority given to consumer demand.

- System performance during standard pricing time interval [10:00, 18:00)

During this standard pricing time interval, the consumer's power requirements and the rate imposed on the electricity by the utility are reasonable. Therefore, consumer demand is exclusively met by the utility which is simultaneously used to increase the SoC of the storage system; these are noticeable from Figs. 10.6d and 10.7c. Figure 10.6b, c corroborates the fact that the photovoltaic and the battery systems do not provide any power to the consumer; this power is fed to the utility to generate income as pointed out in Fig. 10.7d.

- System performance during peak pricing time interval [18:00, 20:00)

In this pricing time interval, the consumer is mainly supplied by the photovoltaic in conjunction with a storage system as illustrated in Fig. 10.6b, c. Figure 10.7d highlights the fact that the available energy stored and not used by the customer is fed into the utility grid to generate substantial income during this peak pricing time interval.

- System performance during standard pricing time interval [20:00, 22:00)

During this standard pricing time interval, it can be seen in Fig. 10.6a that the load is high, reaching a peak of 8 kW. The photovoltaic system does not produce any power at night-time; therefore, the customer's demand is supplied by the battery with a contribution from the when needed as illustrated in Fig. 10.6c, d. The balance of energy from the storage system not used by the customer is injected into the utility grid as shown in Fig. 10.7d.

- System performance during off-peak pricing time interval (22:00, 24:00]

Figure 10.7c illustrates that during this off-peak pricing time interval, the utility is the principal supplier; however, a minimal output from the storage system to the consumer is also noticeable as illustrated in Fig. 10.6d. Figure 10.7d shows the profile of power injected into the grid during this off-peak pricing time interval.

10.5.2 Residential Demand Supplied by the Proposed System

Even though the commercial load profile is different from the residential one, the behaviour and power flow of the hybrid system in the first off-peak pricing period [00:00, 06:00), in the standard pricing period [06:00, 07:00) and in the peak pricing period [07:00, 10:00) are similar for both sectors.

The behaviour and power flow of the hybrid system supplying the commercial load in the standard pricing period [10:00, 18:00) is characterized by the fact that the load demand is high; the grid is used as the main supply to the demand together with the storage system, as illustrated in Figs. 10.8d and 10.9c. However, the photovoltaic system also contributes to supplying the demand to a minor instance, as illustrated in Fig. 10.8b, which was not needed in the residential case. The battery does not provide any power to the consumer, as illustrated in Fig. 10.8c; Fig. 10.8d reveals that there is no power fed into the utility.

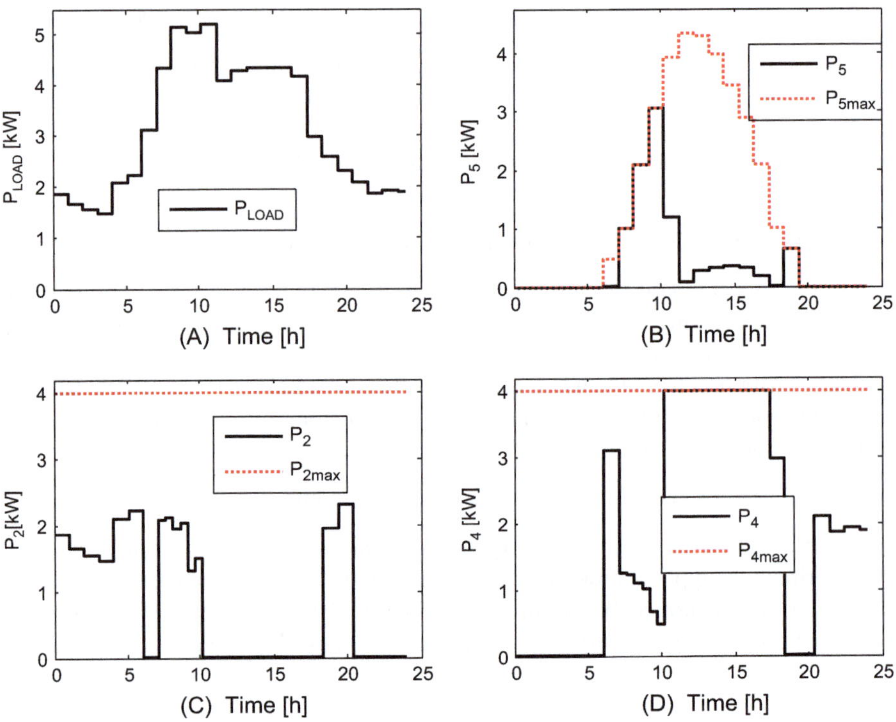

Fig. 10.8 Commercial case: Demand side power scheduling

The performance of the hybrid system supplying the commercial load during the peak pricing period [18:00, 20:00), during the standard pricing period [20:00, 22:00) and during the off-peak pricing period (22:00, 24:00] is similar for the residential ones.

10.5.3 Industrial Demand Supplied by the Proposed System

The behaviour and power flow of the hybrid system supplying the industrial load are very similar to that of the commercial. However, the main difference is that in peak pricing periods [07:00, 10:00), the photovoltaic system is used as main source to respond to the demand requirement in conjunction with a minoroutput

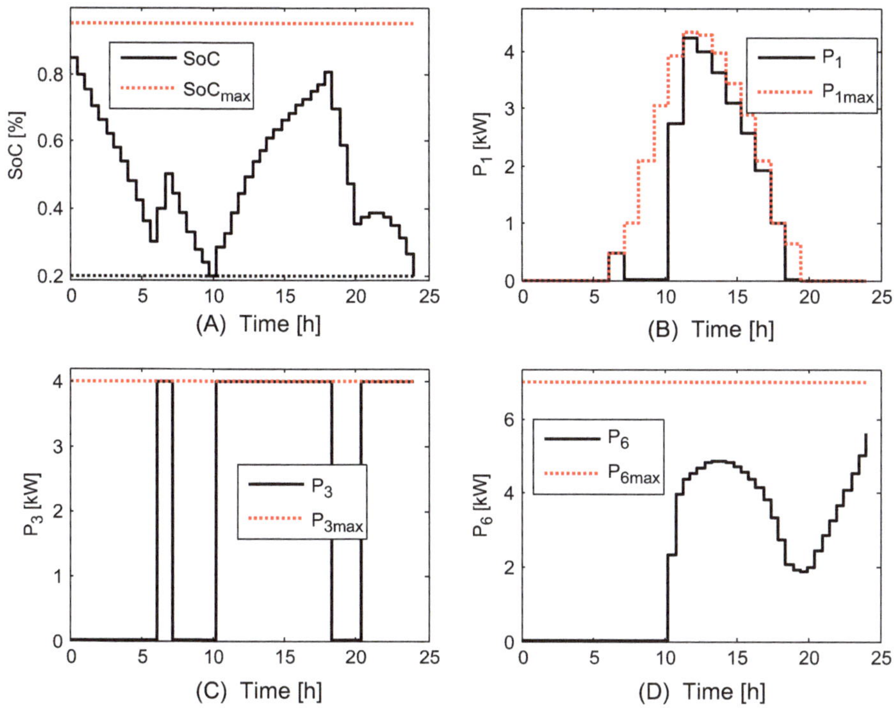

Fig. 10.9 Commercial case: Storage side power scheduling

from the storage system (Fig. 10.10b, c). Specifically, in this case, the surplus of energy available from the storage system not used to feed the consumer is injected into the grid, as illustrated in Fig. 10.11d.

10.5.4 Daily Economic Analysis

A summary of the microgrid's daily running expenses is given in Table 10.1 below. The second column gives the daily cost of electricity when the utility exclusively supplies the consumer. The monetary value of the daily power sold to the utility when the microgrid is optimally scheduled is given in the third column. The fourth column displays the net revenue generated by the consumer (which is the difference between column three and column two).

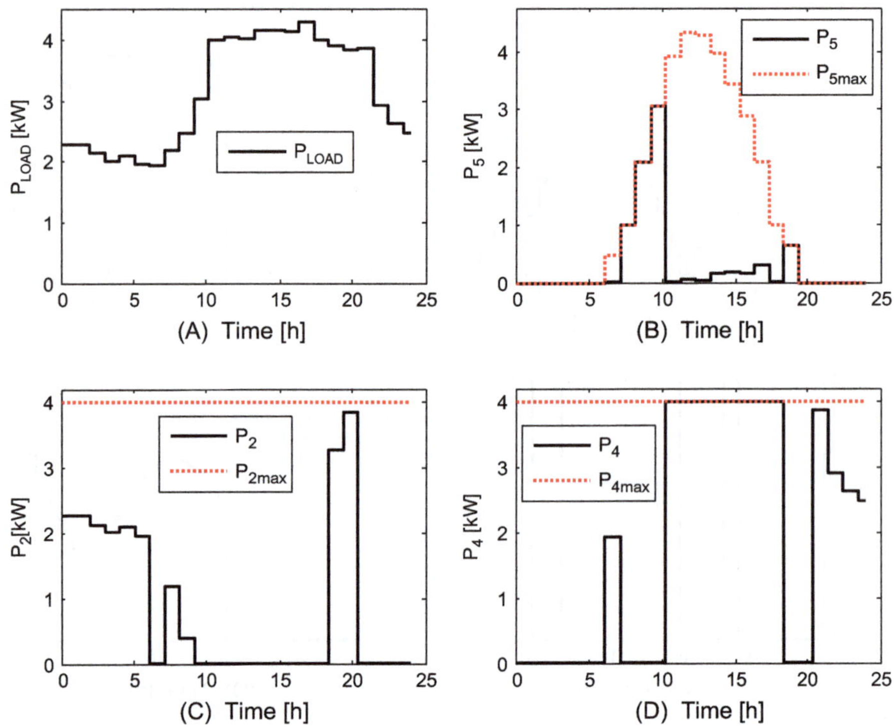

Fig. 10.10 Industrial case: Demand side power scheduling

Even though the energy consumption is the same for the three cases, the results show that the difference in the income generated is mainly a function of the daily load profile. The results show that more saving and income are generated on the industrial and commercial sectors in comparison to the residential sector.

10.6 Chapter Summary

This study focuses on analysing the impact of residential, commercial and industrial load profiles on the daily operational cost and scheduling of grid-connected hybrid systems with the time-of-use tariff in Bloemfontein, South African. Therefore, three profiles from the main demand sectors, namely residential, commercial and industrial load profiles having the same daily energy consumption of 75 kWh, were considered; and techno-economic behaviors of the hybrid system have been conducted regarding to the different hourly pricing periods.

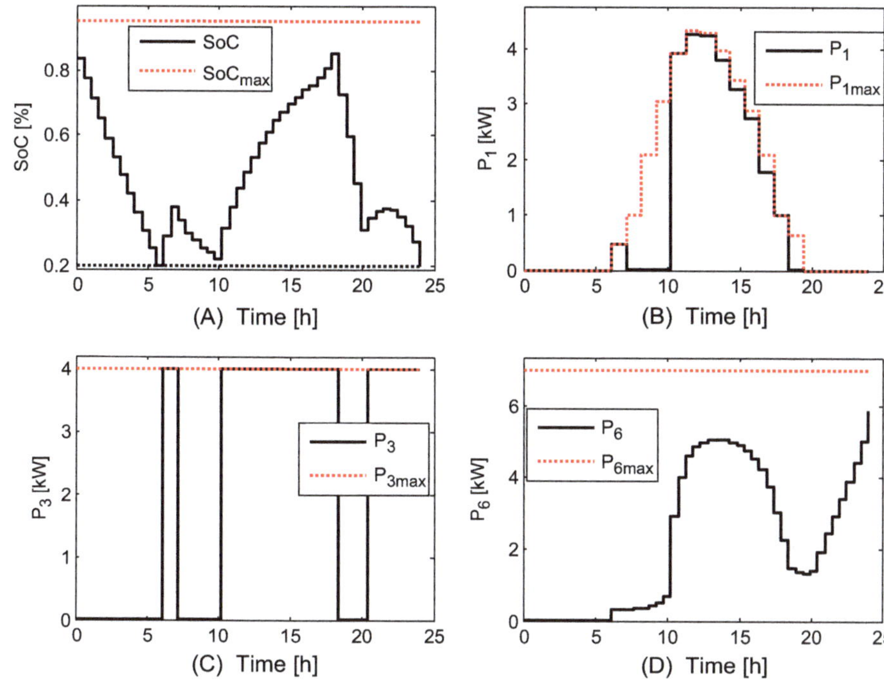

Fig. 10.11 Industrial case: Storage side power scheduling

Table 10.1 Daily operation costs comparison

Load type	Power used: Grid only (R)	Sold: Grid connected hybrid system (R)	Saving (R)
Residential	67.4	−73	−5.6
Commercial	62.4	−85.2	−22.8
Industrial	62.3	−93.9	−31.6

The simulation results have revealed that for the same energy consumption and renewable resources, the daily operational cost of a proposed system is mainly dependent on the demand's profile. It has been shown, using the proposed set-up in this work, that more savings or income is generated from the industrial and commercial sectors than from the residential sector. Therefore, from the results obtained, it can be recommended that in South Africa more focus should be on implementing grid-connected renewable hybrids with storage system in the commercial and industrial sectors rather than in the domestic one.

References

1. Sinha S, Chandel SS. Improving the reliability of photovoltaic-based hybrid power system with battery storage in low wind locations. Sustain Energy Technol Assess. 2017;19:146–59.
2. Kusakana K. Optimization of the daily operation of a hydrokinetic–diesel hybrid system with pumped hydro storage. Energy Convers Manag. 2015;106:901–10.
3. Borisova A, Popov D. An option for the integration of solar photovoltaics into small nuclear power plant with thermal energy storage. Sustain Energy Technol Assess. 2016;18:119–26.
4. Kusakana K. Optimal scheduled power flow for distributed photovoltaic/wind/diesel generators with battery storage system. IET Renew Power Gener. 2015;9(8):916–24.
5. Zubi G, Dufo-López R, Pasaoglu G, Pardo N. Techno-economic assessment of an off-grid PV system for developing regions to provide electricity for basic domestic needs: A 2020–2040 scenario. Appl Energy. 2016;176:309–19.
6. Numbi BP, Malinga SJ. Optimal energy cost and economic analysis of a residential grid-interactive solar PV system-case of eThekwini municipality in South Africa. Appl Energy. 2017;186:28–45.
7. Wua K, Zhoub H, Anb S, Huang T. Optimal coordinate operation control for wind–photovoltaic–battery storage power-generation units. Energy Convers Manag. 2015;90: 466–75.
8. Kusakana K. Optimal operation control of a grid-connected photovoltaic-battery hybrid system. In: IEEE PES power africa conference, June 28–July 3 2016. p. 239–244.
9. Dufo-Lopez R, Bernal-Agustin JL. Techno-economic analysis of grid-connected battery storage. Energy Convers Manag. 2015;91:394–404.
10. Nwulu NI, Xia X. Optimal dispatch for a microgrid incorporating renewables and demand response. Renew Energy. 2017;101:16–28.
11. Abdulaal A, Asfour S. A linear optimization based controller method for real-time load shifting in industrial and commercial buildings. Energy Build. 2016;110:269–83.
12. Ge Y, Zhou C, Hepburn DM. Domestic electricity load modelling by multiple Gaussian functions. Energy Build. 2016;126:455–62.
13. Sichilalu S, Tazvinga H, Xia X. Optimal control of a fuel cell/wind/PV/grid hybrid system with thermal heat pump load. Sol Energy. 2016;135:59–69.
14. Huang C, Edesess M, Bensoussan A, Tsui KL. Performance analysis of a grid-connected upgraded metallurgical grade silicon photovoltaic system. Energies. 2016;9(5):342.
15. Moreno-Garcia IM, Palacios-Garcia EJ, Pallares-Lopez V, Santiago I, Gonzalez-Redondo MJ, Varo-Martinez M, Real-Calvo RJ. Real-time monitoring system for a utility-scale photovoltaic power plant. Sensors. 2016;16(6):770.
16. Kusakana K. Minimum cost solution of isolated battery-integrated diesel generator hybrid systems. In: South African University Power and Energy conference (SAUPEC 2015), 28–30 January 2015. p. 141–147.
17. Kim H, Baek S, Ha Choi K, Kim D, Lee S, Kim D, Chang HJ. Comparative analysis of on- and off-grid electrification: the case of two South Korean islands. Sustainability. 2016;8(4):350.
18. González A, Riba JR, Rius A. Optimal sizing of a hybrid grid-connected photovoltaic–wind–biomass power system. Sustainability. 2015;7(9):12787–806.
19. Orioli A, Franzitta V, Di Gangi A, Foresta F. The recent change in the Italian policies for photovoltaics: effects on the energy demand coverage of grid-connected pv systems installed in urban contexts. Energies. 2016;9(11):944.
20. Kusakana K. Optimal scheduling for distributed hybrid system with pumped hydro storage. Energy Convers Manag. 2016;111:253–60.
21. Mandelli S, Brivio C, Colombo E, Merlo M. Effect of load profile uncertainty on the optimum sizing of off-grid PV systems for rural electrification. Sustain Energy Technol Assess. 2016;18:34–47.
22. Sichilalu SM, Xia X. Optimal power dispatch of a grid tied-battery-photovoltaic system supplying heat pump water heaters. Energy Convers Manag. 2015;102:81–91.
23. Kusakana K. Energy management of a grid-connected hydrokinetic system under Time of Use tariff. Renew Energy. 2017;101:1325–33.

Index

W. Rivera (ed.), *Sustainable Cloud and Energy Services*,
DOI 10.1007/978-3-319-62238-5

C

D

MIX
Papier aus verantwortungsvollen Quellen
Paper from responsible sources
FSC® C105338

If you have any concerns about our products,
you can contact us on
ProductSafety@springernature.com

In case Publisher is established outside the EU,
the EU authorized representative is:
Springer Nature Customer Service Center GmbH
Europaplatz 3, 69115 Heidelberg, Germany

Printed by Libri Plureos GmbH
in Hamburg, Germany